Treatment Technologies for Groundwater

Treatment Technologies for Groundwater

Lee H. Odell

American Water Works
Association

Treatment Technologies for Groundwater
Copyright © 2010 American Water Works Association

Disclaimer

AWWA Publications Manager: Gay Porter De Nileon
Technical Editor/Project Manager: Mary Kay Kozyra
Cover Design/Production Editor: Cheryl Armstrong

Library of Congress Cataloging-in-Publication Data

Odell, Lee H.
  Treatment technologies for groundwater / Lee H. Odell.
    p. cm.
  ISBN 978-1-58321-757-3
  1. Groundwater--Purification. 2. Drinking water--Contamination. I. Title.
  TD462.O34 2010
  628.1'14--dc22

                  2010011846

ISBN 1-58321-757-6
       978-158321-757-3

American Water Works Association

6666 West Quincy Avenue
Denver, CO 80235-3098
303.794.7711
www.awwa.org

Contents

10 Radium and Gross Alpha Removal 171

11 Barium Removal 181

12 Organic Compound Removal 187

Appendix A Materials Compatibility for Chemical Feed Systems 195

Index 199

About the Author 231

List of Figures

List of Tables

Preface

This handbook is a reference for prospective or current operators of groundwater treatment facilities. It is not intended as a design manual; however, it does provide guidance on the important characteristics and criteria to use when selecting, designing, and operating groundwater treatment plants.

This handbook organizes information that pertains specifically to groundwaters used as drinking water supplies in an easy-to-use manner.

CHAPTER ONE

Groundwater Treatment Regulations

In the past, it seemed as if groundwater treatment regulations were an afterthought to surface water regulations. Today, there is a much better understanding of the vulnerability of groundwater supplies to various contaminant threats and a broader regulatory framework that affects groundwater supplies. This chapter includes a review of federal regulations that apply to groundwater used for municipal drinking water supplies and an overview of primary and secondary drinking water standards.

REGULATIONS THAT IMPACT GROUNDWATER SYSTEMS

The US Environmental Protection Agency (USEPA) protects public health by regulating drinking water supplies under the framework of the Safe Drinking Water Act (SDWA) and the Safe Drinking Water Act Reauthorization. Many of these regulations apply to public water systems that use groundwater as a supply. Table 1-1 provides a summary of regulations that apply to groundwater systems and that could result in the requirement to treat groundwater. Full texts of the regulations can be found at http://www.epa.gov/safewater/regs.html.

The USEPA is required to periodically review these regulations and revise them if necessary to maintain or improve the same level of public health protection originally regulated. For example, when the Arsenic Rule was made final in 2001, USEPA stated that the health effects information for the rule was incomplete and that when it was available, the agency would review the health effects information. In 2005, the Science Advisory Board for the USEPA reviewed new health effects information for this rule and made recommendations for a lower maximum contaminant level (MCL). The USEPA is currently reviewing the regulations and incorporating the cost-and-benefit analysis results and the ability of existing technology to treat to lower levels.

Table 1-1   Current USEPA Regulations That Apply to Groundwater Systems

Regulation	Key Provisions for Groundwater	Potential Treatment Implications
Arsenic Rule	Establishes an MCL and an MCLG for arsenic	Treatment to remove arsenic in excess of MCL
Filter Backwash Rule	Systems using groundwater under the influence of surface water and using filtration with recycling of backwash water must recycle in a manner that meets regulatory standards	Potential process changes for recycling of backwash water
Groundwater Rule	Vulnerability standards for groundwater supplies Viral inactivation treatment provisions established for vulnerable systems	Disinfection and contact time for viral inactivation for vulnerable systems
Lead and Copper Rule	Establishes action levels for lead and copper in at-risk customer tap samples	Optimization of corrosion control for lead and copper if ALs exceeded
Long-Term 2 Enhanced Surface Water Treatment Rule	Groundwater under the influence of surface water must treated for *Cryptosporidium*	Disinfection and/or filtration for *Cryptosporidium*
National Primary Drinking Water Standards	Establishes MCLs and some treatment techniques for several synthetic organic chemicals, volatile organic chemicals, inorganic chemicals, physical parameters, and microbial contaminants	Treatment for a variety of compounds if MCLs or ALs are exceeded See this chapter for individual compounds and levels requiring treatment
National Secondary Drinking Water Standards	Establishes secondary MCLs for several inorganics and physical properties of water	Requirements vary by state but are generally viewed as guidelines for drinking water quality
Radionuclide Rule	Establishes MCLs and MCLGs for combined radium-226/228, adjusted gross alpha, beta particle and photon radioactivity, and uranium	Treatment for combined radium-226/228, adjusted gross alpha, beta particle, and photon radioactivity above MCLs

Table 1-1   Current USEPA Regulations That Apply to Groundwater Systems (continued)

Regulation	Key Provisions for Groundwater	Potential Treatment Implications
Stage 2 Disinfectants and Disinfection By-Products Rule	Requires systems to identify and monitor for two classes of disinfection by-products: TTHMs and HAA5 at sites in the distribution system likely to have the highest levels Establishes MCLs for TTHMs and HAA5 at each location tested	Reduction of DBP levels

AL—action level; HAA5—five haloacetic acids; MCL—maximum contaminant level; MCLG—maximum contaminant level goal; TTHM—total trihalomethanes

In addition, the USEPA continues to add new contaminants to the list of Primary Drinking Water Standards. In 2005, the USEPA provided a contaminant candidate list (CCL) that included 8 additional microbials and 42 additional chemicals for possible regulation under the SDWA.

CURRENT PRIMARY AND SECONDARY DRINKING WATER STANDARDS

There are currently drinking water quality standards for 95 contaminants, including 9 microbials, 8 disinfection by-products (DBPs) and residuals, 18 inorganics (including lead and copper), 53 organics, and 7 radiologic contaminants. These standards, which have either established MCLs or identified treatment techniques, are summarized in the following section.

Microbial Contaminants: Coliform Bacteria

Microbial contaminants are regulated for groundwater systems under the following three regulatory mechanisms:

- All groundwater systems are routinely tested for total coliform and, if present, for fecal coliform and *Escherichia coli* (*E. coli*).
- Groundwater systems subject to the Groundwater Rule provisions periodically conduct sanitary surveys, monitor for coliform, and, if significant deficiencies or source water fecal

coliform are found, may need to disinfect for 4-log (99.99%) virus removal.

- Groundwater supplies listed as groundwater under the influence of surface water (GUI) must meet the provisions of the Surface Water Treatment Rule and its successors, the Interim Enhanced Surface Water Treatment Rule, Filter Backwash Rule, Long-Term 1 Enhanced Surface Water Treatment Rule, and Long-Term 2 Enhanced Surface Water Treatment Rule.

Coliform. The presence of total coliforms indicates potential problems with microbial water quality and triggers testing for fecal coliforms and *E. coli*. Fecal coliforms and *E. coli* are bacterial contaminants whose presence indicates that the water may be contaminated with human or animal wastes and that urgent action is required to protect public health, including advising water users to boil their drinking water or use alternate supplies. Microbes in these water supplies can cause short-term health effects such as diarrhea, cramps, nausea, headaches, and other symptoms. They may pose a special health risk for infants, young children, and people with severely compromised immune systems.

Pathogenic organisms. Regulations of specific disease-causing (pathogenic) microbial organisms, including *Cryptosporidium*, *Giardia lamblia*, enteric viruses, and *Legionella*, are usually associated with water systems that use surface water supplies. However, groundwater that is under the influence of surface water may also contain these contaminants. Requirements for microbial contaminants of these pathogenic organisms can also include indicators of microbial contamination, including heterotrophic bacteria (measured by heterotrophic plate count, or HPC) and turbidity.

Pathogenic organisms in drinking water can cause a host of waterborne diseases in humans (Table 1-2). These organisms include bacteria, viruses, and parasites that can cause symptoms such as nausea, cramps, diarrhea, and associated headaches.

GUI systems must provide a total level of treatment to remove/inactivate 99.9% (3-log) of *Giardia lamblia* and to remove/inactivate 99.99% (4-log) of viruses. These systems must also remove or inactivate *Cryptosporidium*. The treatment requirements for *Cryptosporidium* vary based on whether or not the supply is filtered and on how much *Cryptosporidium* is found during source water monitoring. Filtered water systems that recycle spent filter backwash water or other waste flows must return those flows through all treatment processes in the filtration plant.

Table 1-2   Microbial Contaminants

Contaminant	MCL, mg/L	Potential Health Effects	Potential Sources
Cryptosporidium	TT	Gastrointestinal disease	Human and animal fecal wastes
E. coli	Confirmed presence	Most specific indicator of the presence of pathogens	Human and animal fecal wastes
Fecal coliforms	Confirmed presence	More specific indicator of the presence of pathogens	Human and animal fecal wastes, some natural environmental sources
Giardia lamblia	TT	Gastrointestinal disease	Human and animal fecal wastes
Heterotrophic plate count	TT	Indicates water quality, effectiveness of disinfection treatment	Naturally occurring bacteria
Legionella	TT	Legionnaires disease	Natural waters, can grow in water heating systems
Total coliforms	<5% positive*	General indicator of the presence of pathogens	Bacteria naturally present in the environment, human and animal fecal wastes
Turbidity	PS	Interferes with disinfection, indicator of filtration treatment performance	Particulate matter from soil runoff
Viruses	TT	Gastrointestinal disease	Human fecal wastes

*For systems collecting fewer than 40 samples per month, the limit for compliance is no more than one sample per month.

PS—performance standard; TT—treatment technique

Compliance with the regulations for GUI systems can be achieved by meeting one of the three following treatment performance standards:

- meeting filtration and disinfection treatment performance standards for surface water systems,

- meeting disinfection and "natural filtration" standards along with wellhead or source water protection, or
- meeting disinfection treatment standards and exception criteria to remain unfiltered.

Criteria for surface water systems to remain unfiltered have been applied in some cases to GUI sources. The criteria to remain unfiltered relate to source water quality, site-specific issues, performance, and monitoring. These criteria are summarized in Table 1-3.

Disinfectants and Disinfection By-products

DBP regulations are intended to protect public health by limiting the exposure of people to chemical disinfectant residuals and chemical by-products of disinfection treatment. Disinfection treatment that is used to kill microorganisms in drinking water can react with naturally occurring organic and inorganic matter in water to form DBPs. A treatment balance is required to apply levels of disinfection treatment needed to kill pathogenic microorganisms while limiting the levels of DBPs produced. Currently regulated DBPs include total trihalomethanes (TTHMs) and five haloacetic acids (HAA5). Table 1-4 includes the regulatory standards for DBPs and maximum residual levels for disinfectants.

To comply with the current regulations, systems must optimize treatment processes to reduce disinfectant residuals and DBPs. DBPs can be reduced by removing compounds that react with disinfectants and by limiting the residual levels and amount of time the disinfectants are in contact with water. Alternative disinfectants such as chlorine dioxide, ozone, ultraviolet light (UV), and chloramines can reduce TTHM and HAA5 levels while still achieving inactivation of pathogenic organisms. UV, however, is not very effective at disinfecting viruses.

Lead and Copper Regulations

Unlike other regulated contaminants, lead and copper levels are regulated at the customer's tap. Treatment technique requirements are imposed to control lead and copper in drinking water. Lead comes from lead solder and brass fixtures, and copper comes from copper tubing and brass fixtures.

Lead health effects. Infants and young children are typically more vulnerable to lead in drinking water than the general population. Infants and children who drink water containing lead in excess of the action level (AL) could experience delays in their physical or

Table 1-3   Criteria for Surface Water Supplies to Remain Unfiltered*

Criteria	Requirements
Water quality	Less than or equal to 100 total coliform bacteria per 100 mL in 90% of samples collected for a running 6-month period or Less than or equal to 20 fecal coliform bacteria per 100 mL in 90% of samples collected for a running 6-month period No turbidity exceedance of 5 ntu
Site-specific issues	99.9% (3-log) *Giardia* inactivation 99.99% (4-log) enteric virus inactivation 99% (2-log) or 99.9% (3-log) Cryptosporidium inactivation depending on source water quality
Performance	Meet daily disinfection performance standards for virus, *Giardia*, and *Cryptosporidium* inactivation Maintain an approved watershed control program Provide a minimum disinfectant residual of 0.2 mg/L at the entry point to the distribution system Maintain distribution disinfectant residuals in 95% of distribution system samples collected monthly Provide reliable backup equipment Have an annual sanitary survey with no source water quality, disinfection treatment, or watershed control deficiencies Comply with total coliform and disinfection by-products standards Have no history of waterborne disease outbreaks Complete disinfection profiling and benchmarking
Monitoring	Continuous or 4-hr turbidity sampling Source water coliform sampling on any day when source water exceeds 1 ntu Continuous recording of disinfectant residual at entry to distribution system Calculate contact times daily

*Has been applied to groundwater supplies in some cases.
DBP–disinfection by-product; HAA5–five haloacetic acids; MRL–maximum residual level; NOM–natural organic matter; TTHM–total trihalomethanes

mental development. Children could show slight deficits in attention span and learning abilities. Adults who drink contaminated water over many years could develop kidney problems or high blood pressure. The USEPA considers lead a probable human carcinogen.

Copper health effects. Copper is an essential nutrient. However, some people who drink water containing copper in excess of the AL over a relatively short period of time could experience severe gastro-intestinal distress. Some people who drink water containing copper in excess of the AL over many years could suffer liver or kidney damage. Those with Wilson's disease cannot tolerate copper in their systems and should consult their health care provider.

Water systems must target recently built homes with lead-soldered copper plumbing and homes with lead service lines for sample collection. In each sampling round, 90% of samples from homes must have lead levels less than or equal to the AL of 0.015 mg/L and copper levels less than or equal to the AL of 1.3 mg/L. If the ALs are exceeded, the system must conduct periodic public education and either install appropriate treatment, change water sources, or replace plumbing.

Various treatment alternatives are available to reduce lead and copper levels. Their application depends on the source of the lead and copper, i.e., is the lead and copper found in the source water or do they come from materials in the distribution system or customers' plumbing. Most often, corrosion control strategies are used to meet lead and copper regulatory requirements. USEPA has recently revised recommendations for corrosion control optimization. However, new information points to the importance of oxidation–reduction potential conditions in the distribution system at controlling lead levels (see Chapter 3).

Inorganic Contaminants

Inorganic chemicals include certain metals and minerals in drinking water, both naturally occurring and those resulting from agricultural or industrial processes. Inorganic contaminants most often come from the water supply source but can also enter water from contact with materials used for pipes and storage tanks. A new, more stringent drinking water standard was recently established for arsenic and inorganic chemicals. Other existing inorganic chemical levels have been regulated for more than 30 years.

For most inorganic contaminants, health concerns are related to long-term or even lifetime exposures (Table 1-5). Arsenic is a naturally occurring mineral known to cause cancer in humans at high concentrations over years of exposure. Short-term exposure to nitrate and nitrite in infants can interfere with the transfer of oxygen from the lungs to the bloodstream. Infants younger than 6 months who drink water containing nitrate or nitrite in excess of the MCLs could become seriously ill and, if untreated, may die. Symptoms include shortness of breath and blue baby syndrome.

Table 1-4   Regulatory Standards for Disinfectant Residuals and Disinfection By-Products

Contaminant	MCL, mg/L	Potential Health Effects	Common Sources in Drinking Water
Bromate	0.010	Cancer	Ozone reaction with natural bromide in water
Bromodichloromethane	See TTHMs	Cancer; liver, kidney, reproductive effects	Chlorine reaction with NOM
Bromoform	See TTHMs	Cancer; nervous system, liver, kidney effects	Chlorine reaction with NOM
Chloramine	4.0 MRL	Reproductive and developmental effects	Added to water as a disinfectant
Chlorine	4.0 MRL	Reproductive and developmental effects	Added to water as a disinfectant
Chlorine dioxide	0.8 MRL	Reproductive and developmental effects	Added to water as a disinfectant
Chlorite	1.0	Oxidative effects on red blood cells	Chlorine dioxide by-products
Chloroform	See TTHMs	Cancer; liver, kidney, reproductive effects	Chlorine reaction with NOM
Dibromoacetic acid	See HAA5	Cancer; reproductive, developmental effects	Chlorine reaction with NOM
Dibromochloromethane	See TTHMs	Nervous system, liver, kidney, reproductive effects	Chlorine reaction with NOM
Dichloroacetic acid	See HAA5	Cancer; reproductive, developmental effects	Chlorine reaction with NOM
HAA5*	0.060	Cancer and other effects	Drinking water chlorination by-products

(continued)

Table 1-4   Regulatory Standards for Disinfectant Residuals and Disinfection By-Products (continued)

Contaminant	MCL, mg/L	Potential Health Effects	Common Sources in Drinking Water
Monobromoacetic acid	See HAA5	Cancer; reproductive, developmental effects	Chlorine reaction with NOM
Monochloroacetic acid	See HAA5	Cancer; reproductive, developmental effects	Chlorine reaction with NOM
Total organic carbon	Treatment technique (if source water exceeds 2.0 mg/L)	None; used as a surrogate for DBP formation potential	NOM present in surface waters
Trichloroacetic acid	See HAA5	Liver, kidney, spleen, developmental effects	Drinking water chlorination by-products
TTHMs†	0.080	Liver, kidney, central nervous system effects; increased risk of cancer	Drinking water chlorination by-products

*Sum of the concentrations of mono-, di-, and trichloroacetic acids and mono- and dibromoacetic acids.

†Sum of the concentrations of chloroform, bromoform, dibromochloromethane, and bromodichloromethane.

Water systems must meet the established MCLs shown in Table 1-5. Systems that exceed one or more MCLs must either install water treatment systems or develop alternate sources of supply. A variety of water treatment processes are available for reducing levels of specific inorganic contaminants in drinking water, including ion exchange and reverse osmosis. See Chapter 2 for specific contaminants and the applicable treatment alternatives.

Organic Chemicals

Organic chemicals are regulated at specific levels based on their individual health effects. Organic contaminants are most often associated with industrial or agricultural activities that affect drinking

Table 1-5   MCLs and Potential Health Effects of Inorganic Contaminants

Contaminant	MCL, mg/L (or as noted)	Potential Health Effects	Common Sources in Drinking Water
Antimony	0.006	Blood cholesterol increase; blood sugar decrease	Discharge from petroleum refineries, fire retardants, ceramics, electronics, solder
Arsenic	0.010	Skin damage; circulatory system effects; increased cancer risk	Erosion of natural deposits of volcanic rocks; runoff from orchards, glass and electronics production wastes
Asbestos	7 million fibers per liter*	Increased risk of developing benign intestinal polyps	Erosion of natural geologic deposits; decay of asbestos-cement water pipes
Barium	2	Increase in blood pressure	Discharge of drilling wastes; discharge from metal refineries; erosion of natural deposits
Beryllium	0.004	Intestinal lesions	Discharge from metal refineries, coal-burning factories, electrical, aerospace, defense industries
Cadmium	0.005	Kidney damage	Corrosion of galvanized pipes; erosion of natural deposits; discharge from metal refineries; runoff from waste batteries and paints
Chromium (total)	0.1	Allergic dermatitis	Discharge from steel and pump mills; erosion of natural deposits
Copper	1.3†, TT	Gastrointestinal distress; people with Wilson's disease cannot tolerate copper	Plumbing materials

(continued)

Table 1-5   MCLs and Potential Health Effects of Inorganic Contaminants (continued)

Contaminant	MCL, mg/L (or as noted)	Potential Health Effects	Common Sources in Drinking Water
Cyanide	0.2	Thyroid, nervous system damage	Discharge from steel/metal, plastic, fertilizer factories
Fluoride	4†	Bone disease; mottled teeth	Erosion of natural deposits; discharge from fertilizer and aluminum industries; drinking water additive promoting strong teeth
Lead	0.015,† TT	Physical and mental development; kidney function; increase in blood pressure; probable human carcinogen	Plumbing and distribution system materials
Mercury (total inorganic)	0.002	Kidney damage	Erosion of natural deposits; discharges from refineries and factories; runoff from landfills, cropland
Nitrate (as N)	10	Methemoglobinemia (blue-baby syndrome) in infants younger than 6 months	Runoff from fertilizers; leaching from septic tank/drain fields; erosion of natural deposits
Nitrite	1	Methemoglobinemia (blue-baby syndrome) in infants younger than 6 months	Runoff from fertilizers; leaching from septic tank/drain fields; erosion of natural deposits (rapidly converted to nitrate)
Selenium	0.05	Hair and nail loss; numbness in fingers and toes; circulatory problems	Discharge from petroleum and metal refineries; erosion of natural deposits; discharge from mines

Table 1-5   MCLs and Potential Health Effects of Inorganic Contaminants (continued)

Contaminant	MCL, mg/L (or as noted)	Potential Health Effects	Common Sources in Drinking Water
Thallium	0.002	Hair loss; blood changes; kidney, lever, intestinal effects	Leaching from ore-processing sites; discharge from electronics, pharmaceutical products, glass factories

* Greater that 10 µm fiber size.

† Less than 90% of samples in targeted sampling.

‡ A secondary standard is set at 2.0 mg/L.

water sources. Major types of organic contaminants are volatile organic chemicals (VOCs) and synthetic organic chemicals (SOCs) and include industrial and commercial solvents and chemicals and pesticides used in agriculture and landscaping. Organic contaminants can also enter drinking water when materials such as pipes, valves, and paints and coatings used inside water storage tanks come in contact with the water. Health concerns are related to long-term or even lifetime exposures to low levels of contaminant (Table 1-6).

Groundwater systems must meet the established MCLs for organic chemicals. Water supplies that exceed one or more MCL must either install treatment systems or develop alternate water sources. A variety of water treatment processes are available for reducing levels of specific organic contaminants in drinking water, including activated carbon and aeration (see Chapter 2 for a complete list).

Radiologic Contaminants

Radiologic contaminants, both natural and man-made, are *regulated to limit* exposure from drinking water (Table 1-7). Rules were recently revised to include a new MCL for uranium and to clarify and modify monitoring requirements.

The primary health effect from long-term exposure to radionuclide compounds is increased cancer risk. If a water supply exceeds the MCL for radionuclides, the system must either install treatment or develop alternate water sources. A variety of treatment processes are used to reduce radiologic contaminants, including adsorption, ion exchange, and reverse osmosis.

Table 1-6   MCLs and Potential Health Effects of Organic Contaminants

Contaminant	MCL, mg/L (or as noted)	Potential Health Effects	Common Sources in Drinking Water
Acrylamide	TT*	Central nervous system and blood effects; increased risk of cancer	Added to water during water and sewage treatment
Alachlor	0.002	Eye, liver, kidney, spleen effects; anemia; increased risk of cancer	Runoff from herbicides used on row crops
Atrazine	0.003	Cardiovascular, reproductive effects	Runoff from herbicides used on row crops
Benzene	0.005	Decrease in blood platelets; anemia; increased risk of cancer	Discharge from factories; leaching from landfills and gas storage tanks
Benzo(a)pyrene	0.0002	Reproductive difficulties; increased risk of cancer	Leaching from linings of water storage tanks and water pipes
Carbofuran	0.04	Blood, nervous system, reproductive system effects	Leaching of soil fumigant used on rice and alfalfa
Carbon tetrachloride	0.005	Liver effects; increased risk of cancer	Discharge from chemical plants, other industrial activities
Chlordane	0.002	Liver and nervous system effects; increased risk of cancer	Residue of banned termiticide
Chlorobenzene	0.1	Kidney and liver effects	Discharge from chemical and agricultural chemical factories
2,4-D	0.07	Liver, adrenal gland, kidney damage	Runoff from herbicides used on row crops

Table 1-6   MCLs and Potential Health Effects of Organic Contaminants (continued)

Contaminant	MCL, mg/L (or as noted)	Potential Health Effects	Common Sources in Drinking Water
Dalapon	0.2	Minor kidney effects	Runoff from herbicides used on rights-of-way
Dibromo-chloropro-pane (DBCP)	0.0002	Reproductive difficulties; increased risk of cancer	Runoff from soil fumigant used on soybeans, cotton, pineapples, orchards
o-Dichlorobenzene	0.6	Liver, kidney, circulatory system damage	Discharge from industrial chemical factories
p-Dichlorobenzene	0.075	Liver, kidney, spleen damage; anemia; blood effects	Discharge from industrial chemical factories
1,2-Dichloroethane	0.005	Increased risk of cancer	Discharge from industrial chemical factories
1,1-Dichloroethylene	0.007	Liver damage	Discharge from industrial chemical factories
Cis-1,2-Dichloroethylene	0.07	Liver damage	Discharge from industrial chemical factories
trans-1,2-Dichloroethylene	0.1	Liver damage	Discharge from industrial chemical factories
Dichloromethane (methylene chloride)	0.005	Liver damage; increased risk of cancer	Discharge from pharmaceutical and chemical factories
1,2-Dichloropropane	0.005	Increased risk of cancer	Discharge from industrial chemical factories
Di(2-ethylhexyl) adipate	0.4	General toxic and reproductive effects	Discharge from chemical factories

(continued)

Table 1-6 MCLs and Potential Health Effects of Organic Contaminants (continued)

Contaminant	MCL, mg/L (or as noted)	Potential Health Effects	Common Sources in Drinking Water
Di(2-ethylhexyl) phathalate	0.006	Liver effects; reproductive difficulties; increased risk of cancer	Discharge from chemical and rubber factories
Dinoseb	0.007	Reproductive difficulties	Runoff from herbicide used on soybeans and vegetables
Dioxin (2,3,7,8-TCDD)	3×10^{-8}	Reproductive difficulties; increased risk of cancer	Emissions from waste incineration and other combustion; discharge from chemical factories
Diquat	0.02	Cataracts	Runoff from herbicide use
Endothall	0.1	Stomach, intestine effects	Runoff from herbicide use
Endrin	0.002	Liver damage	Residue of banned insecticide
Epichlorohydrin	TT*	Stomach effects; increased risk of cancer	Discharge from industrial chemical factories; impurity in some water treatment chemicals
Ethylbenzene	0.7	Liver, kidney damage	Discharge from petroleum refineries
Ethylene dibromide	0.00005	Liver, stomach, kidney, reproductive system effects; increased risk of cancer	Discharge from petroleum refineries
Glyphosate	0.7	Kidney, reproductive system effects	Runoff from herbicide use
Heptachlor	0.004	Liver damage; increased risk of cancer	Residue of banned termiticide

Table 1-6   MCLs and Potential Health Effects of Organic Contaminants (continued)

Contaminant	MCL, mg/L (or as noted)	Potential Health Effects	Common Sources in Drinking Water
Heptachlor epozide	0.0002	Liver damage; increased risk of cancer	Breakdown of heptachlor
Hexachlorobenzene	0.001	Liver, kidney, reproductive system effects; increased risk of cancer	Discharge from metal refineries, agricultural chemical factories
Hexachloro-cyclopentadiene	0.05	Kidney, stomach damage	Runoff/leaching from insecticide used on lumber, gardens, cattle
Methoxychlor	0.04	Reproductive difficulties	Runoff/leaching from insecticide used on fruits, vegetables, alfalfa, livestock
Oxamyl (Vydate)	0.2	Slight nervous system effects	Runoff/leaching from insecticide used on apples, potatoes, tomatoes
Pentachlorophenol	0.001	Liver and kidney effect; increased risk of cancer	Discharge from wood-preserving operations
Picloram	0.5	Liver damage	Herbicide runoff
Polychlorinated biphenyls	0.0005	Skin, thymus gland, reproductive system, nervous system effects; immune deficiencies; increased risk of cancer	Runoff from landfills; discharge of waste chemicals
Simazene	0.004	Blood effects	Herbicide runoff
Styrene	0.1	Liver, kidney, circulatory system damage	Discharge from rubber, plastic factories; leaching from landfills

(continued)

Table 1-6   MCLs and Potential Health Effects of Organic Contaminants (continued)

Contaminant	MCL, mg/L (or as noted)	Potential Health Effects	Common Sources in Drinking Water
Tetrachloroethylene	0.005	Liver damage; increased risk of cancer	Discharge from factories, dry cleaning
Toluene	1	Liver, kidney, nervous system effects	Discharge from petroleum refineries
Toxaphene	0.003	Kidney, liver, thyroid effects; increased risk of cancer	Runoff/leaching from insecticide used on cattle, cotton
2,4,5-TP (Silvex)	0.05	Liver damage	Residue of banned herbicide
1,2,4-Trichlorobenzene	0.07	Adrenal gland changes	Discharge from textile finishing factories
1,1,1-Trichloroethane	0.2	Liver, nervous system, circulatory system effects	Discharge from metal degreasing sites, other factories
1,1,2-Trichloroethane	0.005	Kidney, liver, immune system damage	Discharge from industrial chemical factories
Trichloroethylene	0.005	Liver damage; increased risk of cancer	Discharge from metal degreasing sites, other factories
Vinyl chloride	0.002	Increased risk of cancer	Leaching from PVC (polyvinyl chloride) pipe; discharge from plastics factories
Xylenes (total)	10	Nervous system damage	Discharge from petroleum, chemical factories

* Treatment technique requirement (limit dosage of polymer treatment chemicals).

Table 1-7 MCLs and Potential Health Effects of Radiologic Contaminants

Contaminant	MCL	Potential Health Effects	Common Sources in Drinking Water
Beta and photon emitters*	4 mrem/yr	Increased risk of cancer	Decay of natural and manmade deposits
Combined radium-226/228†	5 pCi/L	Increased risk of cancer	Erosion of natural deposits
Gross alpha	15 pCi/L	Increased risk of cancer	Erosion of natural deposits
Uranium	30 µg/L	Increased risk of cancer; kidney toxicity	Erosion of natural deposits

* Sampling required only if designated by the primacy agency. Gross beta + photon emitters not to exceed 4 mrem/yr.

† Measured separately.

SECONDARY STANDARDS

Secondary drinking water regulations are nonmandatory water quality standards that have been set for 15 contaminants (Table 1-8). These secondary maximum contaminant levels (SMCLs) are not federally enforceable. They were established as guidelines to assist public water systems in managing the aesthetic qualities of their water, e.g., taste, odor, and color. An exceedance of an SMCL may also result in cosmetic or technical impacts. The SMCLs are not set based on health effects risk.

Aesthetic Effects

Aesthetic effects include tastes, odors, and color. Aluminum, chloride, copper, foaming agents, iron, manganese, pH, sulfate, threshold odor number, total dissolved solids, and zinc standards were set, in part, because of taste and odor or color impacts.

Cosmetic Effects

Skin discoloration is a cosmetic effect related to silver ingestion. This effect, called argyria, does not impair body function. Silver is used

Table 1-8   Secondary Maximum Contaminant Levels

Contaminant	Secondary MCL	Noticeable Effects Above the Secondary MCL	Sources in Drinking Water
Aluminum	0.05–0.2 mg/L	Colored water	Natural or manmade contamination
Chloride	250 mg/L	Salty taste	Natural or manmade contamination; seawater intrusion
Color	15 color units	Visible tint	Natural organic matter; some inorganics such as iron or manganese
Copper	1.0 mg/L	Metallic taste; blue-green staining	Natural contaminant; plumbing materials
Corrosivity	Noncorrosive	Metallic taste; corroded pipes; fixtures staining	Decaying organic matter
Fluoride	2.0 mg/L	Tooth discoloration	Natural or manmade contamination
Foaming agents	0.5 mg/L	Frothy, cloudy; bitter taste; odor	Natural or manmade contamination
Iron	0.3 mg/L	Rusty color; sediment; metallic taste; reddish or orange staining	Natural mineral deposits
Manganese	0.05 mg/L	Black to brown color; black staining; bitter metallic taste	Natural mineral deposits
Odor	3 threshold odor number	"Rotten-egg," musty, or chemical smell	Decaying organic matter
pH	6.5–8.5	Low pH: bitter metallic taste; corrosion High pH: slippery feel; soda taste; deposits	Natural mineral or decaying organic matter
Silver	0.1 mg/L	Skin discoloration; graying of white part of the eye	Natural or manmade contamination
Sulfate	250 mg/L	Salty taste	Natural mineral contaminant

Table 1-8   Secondary Maximum Contaminant Levels (continued)

Contaminant	Secondary MCL	Noticeable Effects Above the Secondary MCL	Sources in Drinking Water
Total dis- solved solids	500 mg/L	Hardness; deposits; colored water; stain- ing; salty taste	Natural mineral contaminants
Zinc	5 mg/L	Metallic taste	Natural or manmade contamination

as an antibacterial agent in many home water treatment devices. Tooth discoloration and/or pitting are caused by excess fluoride exposures during the formative years prior to teeth eruption in children. The secondary standard of 2.0 mg/L is intended as a guideline for an upper boundary level in areas that have high levels of naturally occurring fluoride. It is not intended as a substitute for the lower concentrations (0.7 to 1.2 mg/L), which have been recommended for systems that add fluoride to their water. The SMCL level was set in order to balance the beneficial effects of protection from tooth decay and the undesirable effects of excessive exposures leading to discoloration.

Technical Effects

Corrosivity, and staining related to corrosion, may affect the aesthetic quality of water and can have significant economic implications. Iron and copper corrosion can stain household fixtures and impart an objectionable metallic taste and red or blue-green color to the water. High levels of copper in chlorinated water can stain hair green. In addition, mineral deposits can build up on the insides of hot water pipes, boilers, and heat exchangers, restricting or even blocking water flow. A variety of treatment technologies are available to treat these secondary contaminants (see Chapter 2 for specific technologies).

REFERENCES

USEPA (US Environmental Protection Agency). 2009. *Code of Federal Regulations*, Title 40, Part 141, National Primary Drinking Water Regulations. 40 FR 59570, Dec. 24, 1975; 44 FR 68641, Nov. 29, 1979; and 69 FR 18803, Apr. 9, 2004.

USEPA. 2009. *Code of Federal Regulations,* Title 40, Part 143, National Secondary Drinking Water Regulations. 44 FR 42198, July 19, 1979; 51 FR 11412, Apr. 2, 1986; and 56 FR 3597, Jan. 30, 1991.

CHAPTER TWO

Treatment Technology Overview*

This chapter provides a general overview of treatment technologies that can be used to treat groundwater. For each type of treatment discussed, the following questions are answered:

- How does this treatment technology work?
- What types of treatment issues can this technology effectively address?
- What are the key design requirements?
- What types of residuals are associated with this technology?
- How difficult is it to operate and maintain the technology?
- How do commercially available systems differ?

An overview of the treatment technologies is provided in Table 2-1. Subsequent chapters provide a more detailed discussion of some of the treatment technologies used to remove specific contaminants.

COAGULATION–FILTRATION TREATMENT TECHNOLOGIES

Coagulation in combination with filtration, which is the most widely used technology for treating surface water supplies for turbidity and microbial contaminants, may not be appropriate for many groundwater treatment applications. Recent advances in monitoring and control devices have made it possible for a single operator to monitor and operate several small water systems within a given area, making this type of treatment more applicable to groundwater systems that have wells scattered throughout the distribution system.

---

*Treatment technologies are rapidly changing and evolving, and new applications of treatment technologies may have developed since this handbook was printed.

23

Table 2-1 Treatment Technology Summary

Primary Contaminants	Filter: Coagulation–filtration	Biological Filtration	Hydrous Manganese Oxide Filtration	Oxidation/precipitation/filtration	Membrane: Reverse Osmosis	Nanofiltration	Ultrafiltration	Microfiltration	Sorption: Iron Oxides	Manganese Dioxide	Granular Activated Carbon	Ion Exch: Cation Exchange	Anion Exchange	Electrodialysis Reversal	Precip: Barium Sulfate Precipitation	Excess Lime Softening	Pellet Softening	Aeration	Chem Feed: Ozonation	Permanganate	Chlorine	Chlorine Dioxide	Chloramine	Ultraviolet	UV–Peroxide	Lime	Carbon Dioxide	Soda Ash	Caustic	Silicate	Polyphosphate	Orthophosphate	Limestone
Microbial																																	
Giardia lamblia	✓				✓									✓					✓	✓	✓	✓		✓	✓								
Cryptosporidium	✓	✓			✓	✓	✓	✓						✓							✓	✓		✓	✓								
Legionella	✓	✓			✓	✓	✓	✓						✓					✓	✓	✓	✓	✓		✓								
Heterotrophic plate count	✓			✓	✓	✓	✓	✓						✓									✓	✓	✓								
Turbidity	✓				✓	✓	✓	✓						✓																			
Viruses	✓			✓	✓	✓	✓	✓						✓					✓		✓	✓	✓		✓								
Total coliforms	✓				✓	✓	✓	✓						✓					✓		✓	✓	✓		✓								
Fecal coliforms	✓				✓	✓	✓	✓						✓					✓		✓	✓	✓		✓								
E. coli	✓				✓	✓	✓	✓						✓					✓		✓	✓	✓		✓								

Table 2-1   Treatment Technology Summary

Disinfectants and Disinfection By-products	Filter				Membrane Processes				Sorption			Ion Exchange			Precipitation				Chemical Feed Systems														
	Coagulation–filtration	Biological Filtration	Hydrous Manganese Oxide Filtration	Oxidation/precipitation/filtration	Reverse Osmosis	Nanofiltration	Ultrafiltration	Microfiltration	Iron Oxides	Manganese Dioxide	Granular Activated Carbon	Cation Exchange	Anion Exchange	Electrodialysis Reversal	Barium Sulfate Precipitation	Excess Lime Softening	Pellet Softening	Aeration	Ozonation	Permanganate	Chlorine	Chlorine Dioxide	Chloramine	Ultraviolet	UV–Peroxide	Lime	Carbon Dioxide	Soda Ash	Caustic	Silicate	Polyphosphate	Orthophosphate	Limestone
Bromate	✓	✓			✓	✓	✓	✓			✓			✓					✓	✓	✓	✓	✓	✓	✓								
Bromodichloromethane	✓	✓			✓	✓	✓	✓			✓			✓					✓	✓	✓	✓	✓	✓	✓								
Bromoform	✓	✓			✓	✓	✓	✓			✓			✓					✓	✓	✓	✓	✓	✓	✓								
Chlorite	✓	✓			✓	✓	✓	✓			✓			✓					✓	✓	✓	✓	✓	✓	✓								
Chloroform	✓	✓			✓	✓	✓	✓			✓			✓					✓	✓	✓	✓	✓	✓	✓								
Dibromochloromethane	✓	✓			✓	✓	✓	✓			✓			✓					✓	✓	✓	✓	✓	✓	✓								
Dichloroacetic acid	✓	✓			✓	✓	✓	✓			✓			✓						✓	✓	✓	✓	✓	✓								
Haloacetic acids*	✓	✓			✓	✓	✓	✓			✓			✓						✓	✓	✓	✓	✓	✓								
Trichloroacetic acid	✓	✓			✓	✓	✓	✓			✓			✓					✓	✓	✓	✓	✓	✓	✓								
Total trihalomethanes†	✓	✓			✓	✓	✓	✓			✓			✓					✓	✓	✓	✓	✓	✓	✓								

(continued)

Table 2-1   Treatment Technology Summary (continued)

	Coagulation–filtration	Biological Filtration	Hydrous Manganese Oxide Filtration	Oxidation/precipitation/filtration	Reverse Osmosis	Nanofiltration	Ultrafiltration	Microfiltration	Iron Oxides	Manganese Dioxide	Granular Activated Carbon	Cation Exchange	Anion Exchange	Electrodialysis Reversal	Barium Sulfate Precipitation	Excess Lime Softening	Pellet Softening	Aeration	Ozonation	Permanganate	Chlorine	Chlorine Dioxide	Chloramine	Ultraviolet	UV–Peroxide	Lime	Carbon Dioxide	Soda Ash	Caustic	Silicate	Polyphosphate	Orthophosphate	Limestone
	Filter				*Membrane Processes*				*Sorption*			*Ion Exchange*			*Precipitation*											*Chemical Feed Systems*							
Total organic carbon	✓	✓			✓	✓	✓	✓			✓			✓					✓	✓	✓	✓	✓	✓	✓								
Lead and copper																✓																	
Lead					✓									✓												✓	✓	✓	✓	✓	✓	✓	✓
Copper					✓									✓							✓					✓	✓	✓	✓	✓	✓	✓	✓
Inorganics																																	
Antimony	✓				✓									✓																			
Arsenic	✓	✓	✓		✓	✓	✓	✓	✓				✓	✓		✓																	
Barium					✓									✓																			
Beryllium					✓									✓																			
Cadmium					✓									✓																			
Chromium					✓						✓			✓																			
Cyanide					✓									✓																			

Table 2-1   Treatment Technology Summary

	Filter				Membrane Processes				Sorption			Ion Exchange			Precipitation											Chemical Feed Systems							
	Coagulation–filtration	Biological Filtration	Hydrous Manganese Oxide Filtration	Oxidation/precipitation/filtration	Reverse Osmosis	Nanofiltration	Ultrafiltration	Microfiltration	Iron Oxides	Manganese Dioxide	Granular Activated Carbon	Cation Exchange	Anion Exchange	Electrodialysis Reversal	Barium Sulfate Precipitation	Excess Lime Softening	Pellet Softening	Aeration	Ozonation	Permanganate	Chlorine	Chlorine Dioxide	Chloramine	Ultraviolet	UV–Peroxide	Lime	Carbon Dioxide	Soda Ash	Caustic	Silicate	Polyphosphate	Orthophosphate	Limestone
Fluoride					✓								✓	✓				✓								✓	✓	✓	✓	✓		✓	✓
Mercury					✓									✓												✓	✓	✓	✓	✓		✓	✓
Nickel					✓									✓																			
Nitrate		✓			✓	✓							✓	✓	✓																		
Nitrite		✓			✓									✓	✓																		
Selenium					✓								✓	✓																			
Thallium					✓									✓																			
Organics																																	
Volatile organics											✓							✓	✓	✓	✓	✓			✓								
Synthetic organics	✓	✓			✓	✓					✓		✓	✓		✓			✓						✓								
Pesticides	✓	✓			✓	✓					✓			✓		✓			✓						✓								
Dissolved organic carbon	✓	✓			✓	✓					✓		✓	✓					✓						✓								

(continued)

Table 2-1   Treatment Technology Summary (continued)

	Filter				Membrane Processes				Sorption			Ion Exchange			Precipitation				Chemical Feed Systems														
	Coagulation–filtration	Biological Filtration	Hydrous Manganese Oxide Filtration	Oxidation/precipitation/filtration	Reverse Osmosis	Nanofiltration	Ultrafiltration	Microfiltration	Iron Oxides	Manganese Dioxide	Granular Activated Carbon	Cation Exchange	Anion Exchange	Electrodialysis Reversal	Barium Sulfate Precipitation	Excess Lime Softening	Pellet Softening	Aeration	Ozonation	Permanganate	Chlorine	Chlorine Dioxide	Chloramine	Ultraviolet	UV–Peroxide	Lime	Carbon Dioxide	Soda Ash	Caustic	Silicate	Polyphosphate	Orthophosphate	Limestone
Disinfection by-product precursors	✓	✓			✓	✓					✓			✓		✓			✓						✓								
Pharmaceuticals	✓	✓			✓	✓					✓			✓		✓			✓						✓								
Radionuclides																																	
Gross alpha					✓						✓			✓	✓																		
Beta and photon emitters*					✓						✓			✓	✓																		
Iodine-131†					✓						✓			✓	✓																		
Combined radium-226/228‡			✓		✓					✓	✓			✓	✓																		
Uranium	✓		✓		✓						✓			✓	✓																		
Strontium 90†	✓	✓	✓		✓						✓			✓	✓	✓		✓															
Tritium†	✓	✓	✓		✓						✓			✓		✓																	

Table 2-1   Treatment Technology Summary

Contaminant	Filter				Membrane Processes				Sorption			Ion Exchange			Precipitation				Chemical Feed Systems															
	Coagulation–filtration	Biological Filtration	Hydrous Manganese Oxide Filtration	Oxidation/precipitation/filtration	Reverse Osmosis	Nanofiltration	Ultrafiltration	Microfiltration	Iron Oxides	Manganese Dioxide	Granular Activated Carbon	Cation Exchange	Anion Exchange	Electrodialysis Reversal	Barium Sulfate Precipitation	Excess Lime Softening	Pellet Softening	Aeration	Ozonation	Permanganate	Chlorine	Chlorine Dioxide	Chloramine	Ultraviolet	UV–Peroxide	Lime	Carbon Dioxide	Soda Ash	Caustic	Silicate	Polyphosphate	Orthophosphate	Limestone
Radon	✓	✓	✓		✓						✓			✓		✓																	
Secondary Contaminants																																	
Hardness					✓	✓						✓		✓		✓	✓																
Iron	✓	✓	✓	✓	✓	✓	✓	✓		✓		✓		✓		✓														✓	✓		
Manganese	✓	✓	✓	✓	✓	✓	✓	✓		✓		✓		✓		✓														✓	✓		
Total dissolved solids					✓	✓								✓		✓																	
Chloride					✓									✓		✓																	
Sulfate					✓								✓	✓		✓																	
Zinc					✓	✓				✓				✓		✓																	
Color	✓	✓	✓	✓	✓						✓		✓	✓		✓			✓	✓	✓	✓			✓								
Taste and odor	✓	✓	✓	✓	✓						✓			✓		✓			✓	✓	✓	✓	✓	✓	✓								
Aluminum			✓		✓								✓	✓		✓																	

(continued)

Table 2-1   Treatment Technology Summary (continued)

	Filter				Membrane Processes				Sorption			Ion Exchange			Precipitation											Chemical Feed Systems							
	Coagulation–filtration	Biological Filtration	Hydrous Manganese Oxide Filtration	Oxidation/precipitation/filtration	Reverse Osmosis	Nanofiltration	Ultrafiltration	Microfiltration	Iron Oxides	Manganese Dioxide	Granular Activated Carbon	Cation Exchange	Anion Exchange	Electrodialysis Reversal	Barium Sulfate Precipitation	Excess Lime Softening	Pellet Softening	Aeration	Ozonation	Permanganate	Chlorine	Chlorine Dioxide	Chloramine	Ultraviolet	UV–Peroxide	Lime	Carbon Dioxide	Soda Ash	Caustic	Silicate	Polyphosphate	Orthophosphate	Limestone
Copper					✓									✓												✓	✓	✓	✓	✓	✓	✓	✓
Corrosivity					✓											✓																	
Fluoride		✓												✓		✓																	
Foaming agents											✓								✓						✓								
Odor	✓	✓																✓	✓	✓	✓	✓			✓								
pH																										✓	✓	✓	✓	✓			✓
Silver					✓									✓		✓																	

*Includes 168 individual beta particle and photon emitters.

†Included in beta particle and photon emitters.

‡Maximum contaminant level is established for combined radium 226 and 228.

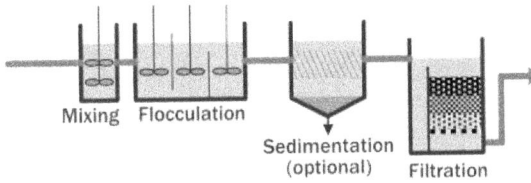

Figure 2-1   Coagulation/filtration process flow diagram
Courtesy of Paul Mueller, CH2M HILL

✤✤ How does coagulation–filtration work?

A process flow diagram for a coagulation–filtration system is shown in Figure 2-1. Coagulation–filtration includes the following pretreatment steps: rapid mixing, chemical coagulation, and flocculation to form settleable or filterable floc particles. Settling is included in some systems that use sedimentation basins, plate settlers, ballasted floc removal, or dissolved air flotation to remove most of the floc particles. The water is then filtered to remove the remaining particles. Coagulation and formation of floc particles are needed in this type of system because the filter media is 500 to 1,000 times larger than the particles being removed. Filter media sizes typically range from 0.4 to 1.6 mm, while particles being filtered are often 1 to 5 µm in size. Common filter media include sand and dual-media (sand and anthracite). Recent trends in coagulation–filtration have been to include deeper media beds and higher filter loading rates. High-rate clarification processes have also been installed more frequently, especially for challenging treatment applications.

✤✤ What types of treatment issues can coagulation–filtration effectively address?

Coagulation–filtration is used to treat a variety of compounds in groundwater. It is a reliable treatment technique for microbial contaminants, arsenic, color, total organic carbon (TOC), and iron and manganese and for treating groundwater under the direct influence of surface water. Optimal coagulant types, doses, and coagulation pH vary depending on water quality and what contaminant (or contaminants) is being removed.

Coagulation–filtration for groundwater treatment is often applied in pressure-filter applications that do not break head. These systems are commonly used for single-wellhead treatment applications to remove inorganics (such as iron, manganese, and arsenic). However, at least one coagulation–pressure filtration system was recently approved for use in California for a well under the direct influence of surface water (California Water Service Company).

❖ What are the key design requirements for coagulation–filtration?

Design criteria are influenced by site-specific conditions; and individual components of the treatment train often vary between systems. Recent trends have been toward deeper bed filters (48 to 72 in. [1.2 to 1.8 m]) with high loading rates (10 to 15 gpm/sq ft [24 to 37 m/hr]). However, some states require filter loading rates at a maximum of 3.0 gpm/sq ft (7 m/hr); this is in accordance with the Recommended Standards for Water Works (Great Lakes–Upper Mississippi 2003). Often states will allow higher filter rates with pilot testing under an approved testing protocol.

Process modifications are often required to optimize contaminant removal for groundwater applications. For example,

- For arsenic removal, prechlorination is often needed to convert arsenic from its reduced form, As(III), to its oxidized form, As(V), prior to coagulation with ferric or alum.
- Preoxidation with a variety of oxidants may be needed to remove color.
- TOC removal may require preoxidation, cationic polymer addition, or low-pH coagulation.
- Iron removal may require preoxidation and possibly coagulant addition if complexed with organics.
- Manganese removal may require preoxidation and pH adjustment.

Pilot testing is recommended before design completion to identify design criteria and optimize the process.

❖❖❖ What types of residuals are associated with coagulation–filtration?

Selection of coagulation–filtration for groundwater treatment does not simply include an evaluation of whether or not the system can effectively remove the contaminant(s) of concern. The amount, concentration, and form of treatment plant residuals are also important considerations. With coagulation–filtration applications, backwash water can be 1 to 5% of production and difficult to settle and reuse. Solids concentrations in backwash water typically range from hundreds of milligrams per liter to thousands of milligrams per liter. Sedimentation wastes must also be treated if clarification is part of the process. Sedimentation residuals are usually a fraction of 1% of production, and solids concentrations are typically less than 3% of the waste stream.

❖❖❖ How difficult is it to operate and maintain coagulation–filtration systems?

If the water quality is steady, coagulation–filtration processes can be fairly easy to operate and maintain. This is often the case with groundwater supplies. If the water quality varies considerably, these systems may need extensive oversight to ensure removal is effective. Because most states consider this type of treatment to be complex, a higher level of operator certification may be required than for an adsorptive process, for example.

❖❖❖ How do commercially available coagulation–filtration systems differ?

A variety of coagulation–filtration package plants applicable to groundwater systems are available from equipment suppliers. In package plants that use sedimentation, sedimentation usually occurs in tube settlers. Some systems include dissolved-air flotation prior to filtration to remove floc particles. In dual-stage filtration, clarification occurs in a tank or vessel that includes some type of media. Some equipment suppliers refer to this as a roughing filter. Typically, roughing filters are not as versatile as sedimentation or flotation; however, some varieties may perform comparably. The clarified water is then passed through a traditional media filter.

A current list of vendors supplying coagulation–filtration systems can be found at http://sourcebook.awwa.org/.

BIOLOGICAL FILTRATION TREATMENT TECHNOLOGIES

Biological filtration is gaining popularity for many treatment applications because of the benefits it provides, including lower disinfection by-product (DBP) concentrations and stable distribution system water quality. Some biological removal groundwater systems are operating in North America to remove iron or manganese and nitrate (Figure 2-2). Some groundwater treatment plants may unintentionally be using biological removal. For example, iron removal plants that use aeration followed by any type of filtration are likely to provide some biological removal in addition to the oxidation/precipitation process.

Figure 2-2   Biological nitrate removal testing in Glendale, Arizona
Courtesy of CH2M HILL

❖ How does biological filtration work?

A process flow diagram for a biological filtration system is shown in Figure 2-3. Processes may vary considerable from site to site depending on water quality and the target treatment concerns. For example, iron and manganese may be removed in two stages, with oxygen addition in the first stage and pH adjustment and additional oxygen addition in the second stage (Figure 2-4).

Biological filtration includes pretreatment steps of aeration, oxygen addition, or ozonation followed by a biological filtration step. Depending on the type of system and the type of bacteria targeted for growth on the filter media, a nutrient feed may be required. For biological nitrate removal, nitrate-reducing bacteria are grown on the media bed. For efficient nitrate reduction, a carbon source is required to reduce forms of oxidized nitrogen to nitrogen gas. California currently requires a postfiltration step after the biological filtration step to prevent sloughing particles from passing through the underdrain of the biological filter and entering the distribution system.

By contrast, biological removal for iron does not require a carbon source or any food source other than iron. The autotrophic bacteria that remove iron, including stalked bacteria such as *Gallionella ferruginea* and filamentous bacteria such as *Leptothrix ocracea*, absorb the small amount of energy that is given off when iron is changed from its reduced form to its oxidized form and then use this energy to sustain growth.

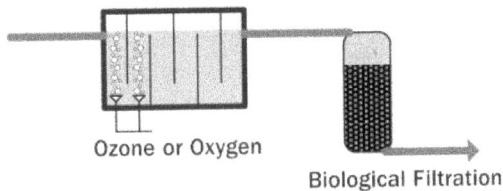

Ozone or Oxygen

Biological Filtration

Figure 2-3 **Biological filtration process flow diagram**
Courtesy of Paul Mueller, CH2M HILL

Figure 2-4   Biological iron and manganese removal vessels in Germany

❖❖❖ What types of treatment issues can biological filtration effectively address?

Theoretically, biological filtration can be used to remove many contaminants, including

- nitrate,
- iron,
- manganese,
- hydrogen sulfide,
- color,
- pharmaceuticals,
- many synthetic organic compounds, and
- many natural organic compounds.

There are few commercially viable systems for many of these applications. However, a significant amount of research is currently under way that may lead to greater availability of biological treatment systems.

❖❖❖ What are the key design requirements for biological filtration?

For biological removal of iron and manganese, oxidation reduction potential (ORP) appears to be the key requirement for abundant growth of iron bacteria. Table 2-2 provides reported ORP conditions for biological removal of iron and manganese.

Table 2-2   Oxidation Reduction Potential Conditions for Biological Removal of Iron and Manganese

	pH 6.0	pH 7.0	pH 8.0	pH 9.0
Iron removal	100–500 mV	50–350 mV	0–130 mV	
Manganese removal			320–570 mV	230–320 mV

Source: Gage et al., 2001.

Biological removal of iron and manganese is often accomplished by growing bacteria on a media bed consisting of granular activated carbon (GAC), greensand, sand, anthracite, or manganese dioxide. The bacteria require considerable time, typically, 4 to 6 weeks under operating conditions, before removal is optimized. Systems are commonly designed with empty bed contact times (EBCTs) of 2.5 to 5 min. Aeration by oxygen addition is used to adjust the water's ORP level, although pH adjustment is often needed to optimize either iron or manganese removal. Pilot testing is recommended.

Biological removal of nitrates can be accomplished in both heterotrophic reactors and autotrophic reactors. However, most existing biological denitrification treatment plants use heterotrophic bacteria. Carbon sources for nitrate removal include sucrose, methanol, and vinegar. Reactor vessels include both packed-bed systems and fluidized-bed systems. Media include cellulose, sand, anthracite, and GAC. EBCTs range from 5 to 20 min. Designs vary with nitrate levels and water quality parameters. Postfiltration is commonly used to remove particles and excess carbon substrate. Pilot testing is recommended to establish design criteria.

❖ What types of residuals are associated with biological filtration?

An important benefit of biological filtration technologies is that it is generally easy to dispose of the plant residuals. The residuals will likely have elevated total suspended solids. However, the suspended solids are usually comprised of nonpathogenic biological growth and can be easily disposed of in sanitary sewers or put to beneficial use. Chlorine is often not a component of concern in backwash residuals for biological plants.

❖❖❖ How difficult is it to operate and maintain biological filtration systems?

Biological filtration plants are generally very easy to operate and maintain, with a few exceptions. The bacteria may take several weeks to grow to a population size that can efficiently remove the contaminant of concern at start-up. Because of long start-up times, careful planning is needed to avoid wasting several weeks' worth of water production during start-up. A strategy to avoid this includes starting up multiple treatment trains, with one train starting in biological mode, while a second train operates using oxidation/precipitation/filtration or adsorption unit processes.

❖❖❖ How do commercially available biological filtration systems differ?

Currently, few biological filter systems are available commercially. However, as their popularity grows, more commercial systems are likely to become available. Systems vary significantly depending on their application. Biological iron and manganese systems may include a single filtration stage or multiple stages that incorporate ozonation, aeration, and pH adjustment. Nitrate removal can be accomplished anaerobically or aerobically and may include reactors and posttreatment processes to remove excess organics and sloughed or washed out bacteria. Design requirements, including those for posttreatment, may also vary from state to state.

A current list of vendors supplying biological filtration systems can be found at http://sourcebook.awwa.org/.

HYDROUS MANGANESE OXIDE FILTRATION

Hydrous manganese oxide (HMO) filtration is an effective and inexpensive process for removing radium from water, especially in cases where a filtration plant already exists.

❖❖❖ How does hydrous manganese oxide filtration work?

A process flow diagram for an HMO filtration system is shown in Figure 2-5. HMO is a precipitated form of manganese that is prepared by mixing manganous sulfate and permanganate. Once the

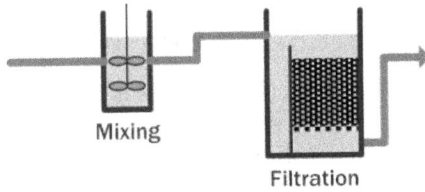

Figure 2-5   Hydrous manganese oxide process flow diagram
Courtesy of Paul Mueller, CH2M HILL

Figure 2-6   Hydrous manganese oxide solution tanks and solution feed pumps
Courtesy of John Dillon, Water Supervisor, City of Batavia, Illinois

mixture is prepared in solution, the freshly precipitated manganese is injected into the water supply using a chemical feed pump (Figure 2-6). Radium quickly sorbs onto the manganese dioxide particles and can be filtered out in traditional media filters or in an iron and manganese filter. Although this is a relatively new technology, there are several installations in the Midwest that are performing well.

❖ What types of treatment issues can hydrous manganese oxide filtration effectively address?

HMO filtration is specifically used to remove radium 226- and -228.

❖❖❖ What are the key design requirements for hydrous manganese oxide filtration?

Important design criteria for this technology are the sizing and specification of the chemical feed system and the design of the filter used to remove the manganese solids. HMO must be freshly formed before use and must be continually mixed to prevent it from settling out in the solution tank. Carrier water systems are often used to carry the HMO to the point of application in order to prevent particles from settling and clogging pipes. Chemical feed pumps must be capable of pumping particulate solutions and must be easy to maintain. Hose (large, peristaltic) pumps work well for this application.

Typical design criteria include filter loading rates of 3 to 5 gpm/sq ft (7 to 10 m/hr) and HMO doses of 0.5 to 2 mg/L. Residuals handling systems are critical for this technology, as is proper media bed design and backwashing systems.

❖❖❖ What types of residuals are associated with hydrous manganese oxide filtration?

Residuals from this process may have elevated radiologic properties. Currently, there is no federal regulation for naturally occurring radioactive wastes (NORM) from water treatment processes. Regulation of these wastes is left to the states for permitting and treatment requirements.

The US Environmental Protection Agency's (USEPA's) Office of Groundwater and Drinking Water published guidelines for the disposal of drinking water treatment wastes containing naturally occurring radionuclides. USEPA guidance suggests disposal to a landfill or a licensed, low-level, radioactive waste disposal facility. When selecting the disposal option, the concentration of radioactive contaminants in the residuals is the governing factor. Unfortunately, no current federally established levels of radionuclides exist to define low or high radioactive wastes or dictate the acceptable disposal method.

Per USEPA, landfill disposal wastes that contain less than 3 pCi/g (dry weight) of radium and less than 50 mg/g of radium may be disposed of in a municipal landfill without the need for long-term institutional controls if the wastes are first dewatered and then spread and mixed with other materials when emplaced. The total contribution of radioactive wastes to the landfill should constitute a small fraction (less than 10% of the volume) of the material in the landfill. Sites that fully comply with USEPA's Subtitle D regulations and guidance under the Resource Conservation and Recovery Act (RCRA) would be appropriate for disposal of this waste.

Methods that comply with USEPA's disposal standards for mill tailings should be considered (40 CFR 192). A decision not to apply these methods fully should be based on a significant difference between the quantity and potential for radium migration on mill tailings versus that on the water treatment plan residual. The disposal method should be augmented by long-term institutional controls to avoid future misuse of disposal sites. At a minimum, disposal in a RCRA-permitted hazardous waste unit should be considered.

At concentrations approaching 2,000 pCi/g, disposal in a licensed low-level radioactive waste disposal facility or facility that is permitted by USEPA or a state for disposal of discrete wastes should be considered. In states where lower-concentration waste disposal is licensed or permitted, that option should be considered for disposal of solids containing 50 to 500 mg/g (dry weight). It is suggested that solid waste containing more than 500 mg/g (dry weight) radium be disposed of in a low-level radioactive waste disposal facility or at a facility that is permitted by USEPA or a state for disposal of NORM wastes.

Expertise is needed to develop safe procedures for handling residuals from HMO plants. In some states, residuals can be discharged to a sanitary sewer or handled on site. Careful calculations are required to ensure that the residuals fall within disposal limits.

❖ How difficult is it to operate and maintain hydrous manganese oxide filtration systems?

Operations and maintenance (O&M) requirements for these facilities include using permanganate and manganous sulfate chemicals to generate HMO chemical feed solutions. The mixing of solutions requires fairly precise measurements and the use of personal protective equipment. Chemical feed facilities must be routinely cleaned and maintained to prevent clogging by HMO particles, which readily precipitate. Solutions of HMO should be freshly precipitated before use and be continually mixed prior to use.

❖❖ How do commercially available hydrous manganese oxide filtration systems differ?

There are no known suppliers of complete HMO package treatment systems. Typically, HMO chemical feed systems are combined with a pressure-filter or gravity-filter system. Permanganate suppliers may be helpful in identifying companies with HMO experience.

OXIDATION/PRECIPITATION/FILTRATION TREATMENT TECHNOLOGY

Oxidation followed by filtration is widely used to treat groundwater. The most common application is oxidation of iron and manganese, which creates a precipitate that can be filtered in a subsequent step. A process flow diagram for an oxidation/precipitation/filtration facility is shown in Figure 2-7.

❖❖ How does oxidation/precipitation/filtration work?

In the oxidation/precipitation/filtration process, the material being removed is first oxidized. Oxidation causes a precipitate to be formed. The precipitated material is then filtered through a media bed. For example, iron dissolved in water is in the form of ferrous iron within the pH range of 6 to 10. To remove iron, an oxidant reacts with the ferrous iron and causes it to precipitate as ferric iron. Once the iron has precipitated, it can be filtered as a particle.

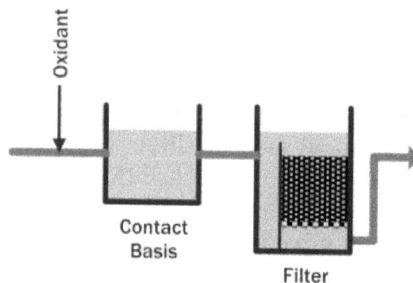

Figure 2-7   Oxidation–filtration process flow diagram

Courtesy of Paul Mueller, CH2M HILL

The most common chemical oxidants used in groundwater treatment are aeration (oxygen), chlorine, ozone, chlorine dioxide, and permanganate (Figures 2-8 and 2-9). Chlorine dioxide can also be used to effectively oxidize manganese, even when high levels of organic material are present. Iron oxidation with chlorine dioxide can be effective, but it is less effective for organically complexed iron compounds.

Figure 2-8   Small on-site sodium hypochlorite generator with wall-mounted reaction cell, softener, brine tank, and solution tank in a well house in California

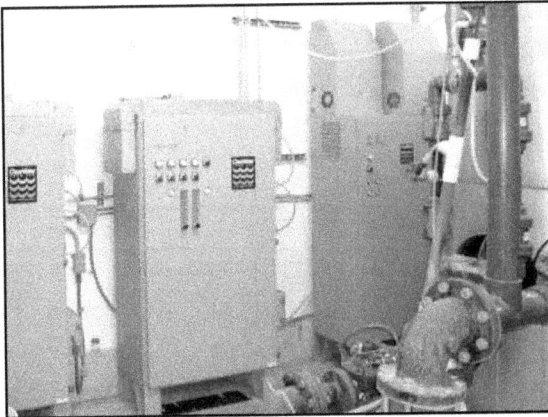

Figure 2-9   Wall-mounted ozone generation equipment in a well building at Camano Island, Washington

Table 2-3 shows the theoretical amount of each oxidant that must be added to completely oxidize 1 mg/L of iron and 1 mg/L of manganese. Additional oxidant may be required to overcome oxidant demands from ammonia or organic compounds.

Determining the amount of oxidant to add is only half of the puzzle when it comes to oxidation. The other half is to determine the amount of time the chemical takes to react with the compound. Table 2-4 shows the reaction times for complete oxidation within the pH range of 6 to 9.

Many reactions are pH dependent, and some reactions, such as the oxidation of manganese with oxygen, simply take too long to make a treatment process design very efficient in many cases.

❖ What types of treatment issues can oxidation/ precipitation/filtration effectively address?

Oxidation/precipitation/filtration processes are usually used to remove iron and manganese. It is possible to remove arsenic using

Table 2-3   Amount of Oxidant Required to Oxidize Iron and Manganese

Oxidant	Per mg/L of Manganese	Per mg/L of Iron
Oxygen (from aeration)	0.29	0.14
Ozone	0.67	0.43
Chlorine	1.28	0.63
Potassium permanganate	1.92	0.94
Chlorine dioxide	2.4	1.2

Table 2-4   Oxidation Reaction Times for Iron and Manganese in Water, pH 6 to 9

Oxidant	Iron Oxidation Rate	Manganese Oxidation Rate
Oxygen (aeration)	<10 min to 4 hr	80 min to 2 days
Ozone	<1 min	< 5 min
Chlorine	Instantaneous to 1 hr	15 min to 12 hr
Permanganate	<5 min	<7 min
Chlorine dioxide	<5 min	<5 min

this process; however, some iron must be present in the water for the arsenic to coprecipitate after oxidation. Hydrogen sulfide can also be oxidized and precipitate as elemental sulfur. Organic compounds can be partially oxidized. However, because DBPs are likely to form, a careful evaluation is needed.

What are the key design requirements for oxidation/precipitation/filtration?

The important design criteria for oxidation/precipitation/filtration include

- identifying the proper type of chemical used for precipitation;
- identifying the dose needed to oxidize the compound to be precipitated, any additional demand in the water, and any desired residual;
- selecting chemical feed equipment that is compatible with the oxidant;
- designing facilities that provide adequate reaction time and/ or adjusting pH to allow oxidation to take place; and
- designing a filter media bed that will effectively remove the particles formed.

Because precipitated iron and manganese particles are normally 1 to 20 μm in size, the filter bed must be carefully designed in order to properly retain these particles in a filter media bed. States often prescribe default filter bed design and filter loading rates. These may require a bed depth of 12 to 24 in. (0.3 to 0.6 m) with 0.45- to 0.55-mm sand topped with 12 to 24 in. (0.3 to 0.6 m) of 0.9- to 1.2-mm anthracite. Loading rates are prescribed at less than 3 gpm/sq ft (7 m/hr) or less than 5 gpm/sq ft (12 m/hr). These "off-the-shelf" filter bed designs often perform poorly.

Many filter bed designs have been developed specifically for iron and manganese particle retention. These are discussed in more detail in Chapter 5.

What types of residuals are associated with oxidation/precipitation/filtration?

The backwash residuals from this treatment process will contain the precipitated contaminants and may contain residual levels from the oxidant. One simple way to calculate the expected residual concentrations is to look at the expected backwash volume as a percentage of production. For example, if a system uses 5% of its water to back-wash, the residuals concentration of the precipitated compound can be estimated as follows:

$$(P_{Co} - P_{Ce})/BW\%$$

Eq. 2-1

where: P_{Co} is the raw water concentration,

P_{Ce} is the finished water concentration, and

BW% is the percent of production volume used for backwashing.

❖ How difficult is it to operate and maintain oxidation/precipitation/filtration systems?

Properly sized and designed systems are generally easy to oper-ate (Figure 2-10). Several treatment issues must be monitored, including

- raw water quality, which may vary after initial start-up or over time;
- chemical feed systems (It is often tempting to reduce chem-ical feed to reduce operating costs. Although performance may seem fine initially, it will likely suffer over time.);
- effluent water quality should be routinely monitored for short-term performance and longer-term trends; and

Figure 2-10 Online process monitoring equipment used to track performance of remote groundwater treatment plants

- backwash duration, frequency, and flow rate should be monitored to make sure the system is being cleaned properly (An entire backwash cycle should be watched at least monthly to ensure the filters are being cleaned properly.).

It is often tempting to reduce backwash duration or flow rate or to extend the time between backwashes to reduce the waste rate. However, many poorly performing systems can be tracked backed to changes made in the backwashing procedure months after they were initially made.

❖ How do commercially available oxidation/precipitation/filtration systems differ?

Oxidation/precipitation/filtration equipment varies significantly from vendor to vendor. These changes are often necessitated by the treatment processes and removal mechanisms that are used. It is important to understand the removal mechanisms used when selecting a potential treatment system.

A current list of vendors supplying oxidation/precipitation/filtration systems can be found at http://sourcebook.awwa.org/.

ADSORPTIVE TREATMENT TECHNOLOGIES

Adsorptive treatment systems have been used for numerous common groundwater treatment applications. New adsorptive media are being developed each year, and the use of this type of treatment technology is expected to continue to grow in popularity and breadth of treatment applications.

❖ How does adsorptive treatment work?

A typical process flow diagram for adsorptive treatment is shown in Figure 2-11. Adsorption works by forming weak bonds between the compound being adsorbed and the media it is adsorbed onto. Commonly used sorption media for municipal drinking water applications include iron oxides, manganese oxides, and GAC.

Iron oxides. Some iron oxide species (e.g., ferric hydroxides) have proven to be good adsorbents for metal ions and some natural organic compounds. A potential benefit of using iron oxides as adsorbents is that their surface charge (both polarity and intensity) can be easily

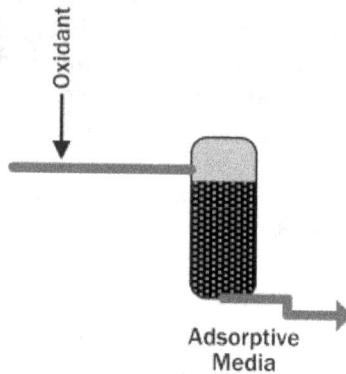

Figure 2-11 Adsorptive removal process flow diagram
Courtesy of Paul Mueller, CH2M HILL

altered by adjusting the solution pH. This special feature allows the use of iron oxides for removal of either cations or anions, depending on solution pH. Also, the used (saturated) iron oxides can be easily regenerated in situ by reversing the solution pH from alkaline to acidic, thereby becoming refreshed and reused. Iron oxides have been most widely used to remove arsenic. In this application, the media is usually not regenerated; rather it is removed, disposed of, and replaced with fresh media.

Manganese dioxide. The dissolved (or reduced) forms iron and manganese can be adsorbed onto manganese dioxide. Adsorption kinetics are much faster than oxidation kinetics. In laboratory tests, Knocke (1990) found that most uptake of manganese at concentrations of up to 1.0 mg/L occurred in the top 6 in. of the media. This finding was also repeated in full-scale plants in Durham, N.C.

Later findings by Knocke et al. (1991) included the following:

- The sorption of Mn(II) by MnOx(s)-coated filter media is very rapid. Both sorption kinetics and sorption capacity increase with increasing pH or surface MnOx concentration.
- In the absence of a filter-applied oxidant, Mn(II) removal is by adsorption alone.
- When free chlorine is present, the oxide surface is continually regenerated, promoting efficient Mn(II) removal over extended periods of time.

Media used for adsorption of manganese and iron include manganese greensand, oxide-coated media, and pyrolusite (manganese dioxide ore) (Figures 2-12 and 2-13).

To maintain efficient uptake kinetics, the oxidative state of the manganese dioxide must be maintained. This can be done by adding permanganate either continuously or periodically. Many applications have been completed without permanganate and often include a continuous application of a free chlorine residual in the range of 0.5 to 1.0 across the media bed.

Granular activated carbon. Although GAC does trap some particles, it works primarily through adsorption, a process in which the organic, radionuclide, or other matter present in water adheres to the carbon granules. GAC can be used to remove color, taste, odors, some radionuclides, and many organic chemicals. The irregular, creviced surface of 1 g of GAC has a surface area of about 600 to 1,000 m^2. Microorganisms also may grow on the surface, feeding on the nutrients in the water and the particles that stick to the carbon.

Critical design features in GAC contactors include the type of GAC and the contact time or length of time the water is in contact with the carbon. Coconut shell GAC is most often used for groundwaters containing low TOC concentrations, hydrogen sulfide, and volatile organic compounds. Bituminous GAC is often used to remove color, high levels of TOC, and large-chain organic compounds. EBCTs may range from 10 to 20 min.

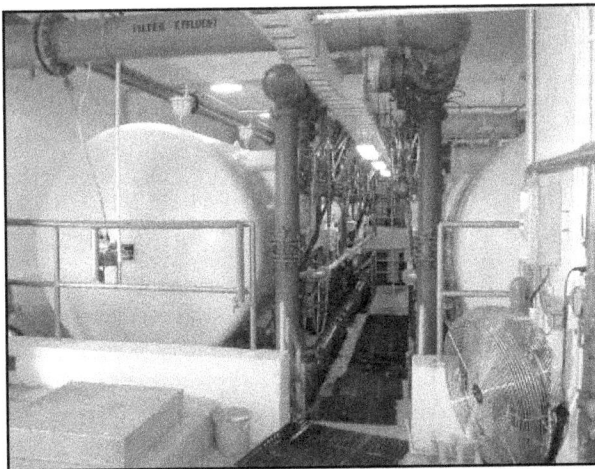

Figure 2-12 Greensand filtration plant in Geneva, Illinois

Figure 2-13 A 4,500-gpm groundwater treatment plant in Batavia, Illinois, that uses manganese dioxide filters and on-site chlorine generation to remove iron and manganese

Photo Courtesy of John Dillon, Water Supervisor, City of Batavia, Illinois

Carbon filters require regular backwashing to clean out buildup of trapped particles; backwashing will not remove any matter that is adsorbed to the carbon. Once the carbon has adsorbed all the organic matter it can, it will be "exhausted," and particles previously adsorbed will then pass through the filter. The carbon must then be replaced or regenerated. Regeneration is accomplished by heating the carbon to high temperatures; this is generally not done on site. Because disinfectant is normally added after the GAC contactor, chlorine will also be removed by the GAC.

❖ What types of treatment issues can adsorptive treatment effectively address?

Iron oxides and hydroxides, aluminum oxides, titanium oxides, and zero-valence iron compounds have been used successfully for arsenic removal and have been tested for removal of other compounds. Manganese dioxide has been widely used to remove iron and manganese, and GAC has been widely applied to remove various inorganic metals, organic compounds, and radionuclides.

Figure 2-14 Iron and manganese filters at Lakewood, Washington, with filter loading rate of 10 gpm/ft^2 using manganese dioxide media

❖❖❖ What are the key design requirements for adsorptive treatment?

Adsorption processes are developed based on adsorption isotherms or pilot testing. Selection of the proper media and sizing the contactor are keys to long-term, effective removal. Some adsorptive media require oxidation for effective continuous removal. In these systems, it is the oxidized state of the media surface that adsorbs the compounds being removed (Figure 2-14).

❖❖❖ What types of residuals are associated with adsorptive treatment?

Residual types vary widely with different types of media. In some cases, the adsorptive media is simply backwashed; the remaining residuals, which have concentrated levels of the contaminant, are then removed. This is typically the case for iron and manganese removal systems. For other adsorptive processes, the contaminants cannot be removed with backwashing; arsenic adsorption to iron hydroxides is an example. GAC is another example. Residuals for these systems often include the entire spent media bed, with the contaminant adhering to the media. Some adsorptive systems can be regenerated with pH adjustment. In these systems, the residual is an alkaline or acid liquid waste with elevated concentrations of the contaminant removed.

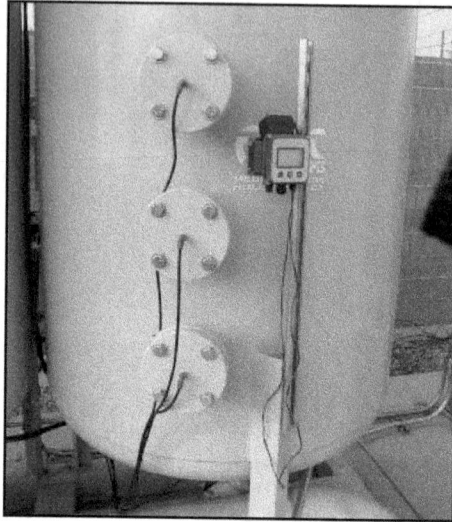

Figure 2-15 Multiple sample ports on adsorptive media vessels help track progression of contaminants through the media bed

❖❖ How difficult is it to operate and maintain adsorptive treatment systems?

Adsorptive systems are relatively easy to operate and maintain. Monitoring is required to ensure that the contaminant is being removed (Figure 2-15). Backwashing must be performed at the proper frequency, duration, and volume, and spent media may need to be replaced periodically. Some systems require pH adjustment or continuous oxidation to remain effective or to optimize performance. In addition, some systems require periodic replacement of media, while others may last for 20 years or more before replacement is needed.

❖❖ How do commercially available adsorptive treatment systems differ?

Adsorptive systems are relatively simple in design but vary from vendor to vendor based on materials, number of valves, backwash requirements (if any), and control systems. A current list of vendors supplying adsorptive removal systems can be found at http://sourcebook.awwa.org/.

ION-EXCHANGE TREATMENT TECHNOLOGY

Ion-exchange treatment has been applied throughout the United States to treat groundwater. Oddly enough, most applications are found in residential water softeners, not municipal treatment systems. However, their application for water treatment extends far beyond the home.

❖❖ How does ion exchange work?

A process flow diagram for an ion-exchange system is shown in Figure 2-16. The system works by loading a resin with an easily displaceable ion. When water passes over the resin, ions in the water exchange places with the displaceable ion. Ion-exchange processes are divided into two types of systems based on the type of ion: systems that remove positively charged ions are cation-exchange systems and systems that remove negatively charged ions are anion-exchange systems.

Cation exchange. A resin with an attraction to positively charged molecules (such as calcium and magnesium) is used for this application. The resin is initially loaded with an exchangeable concentration of a weak cation, e.g., sodium. This cation is then released when positively charged materials are exchanged as they pass over the resin.

The cations on the resin are eventually exhausted and replaced by the cations of the contaminant being removed. When this occurs, the bed must be backwashed, soaked in a regenerant solution (usually the same weak cation that was used in the initial loading), and rinsed, which recharges the bed and removes the built-up contaminant. Cation-exchange resins are either in the sodium form or the hydrogen form. Resins in the sodium form are regenerated with sodium chloride. Potassium chloride can also be used for recharging. Resins in the hydrogen form are regenerated with an acid that has a high concentration of available hydrogen ions.

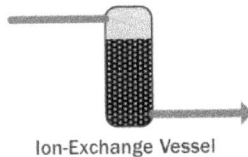

Ion-Exchange Vessel

Figure 2-16 Ion exchange is often a simple flow-through process
Courtesy of Paul Mueller, CH2M HILL

Anion exchange. A resin with an attraction to negatively charged molecules (such as nitrates and sulfates) is used for anion exchange. This resin is initially loaded with a weak cation, e.g., chloride. The chloride is released when negatively charged materials are exchanged as they pass over the resin.

The anions on the resin are eventually exhausted and replaced by the anions of the contaminant being removed. When this occurs, the bed must be backwashed with chloride, which recharges the bed and removes the built-up contaminant. Anion-exchange resins in the chloride form are regenerated with sodium or potassium chloride. Anion-exchange resins in the hydrogen form are regenerated with caustic soda.

❖❖❖ What types of treatment issues can ion exchange effectively address?

Cation exchange is used to remove calcium, magnesium, iron, manganese, and some forms of radionuclides including radium (Figure 2-17). Anion exchange is used to remove fluoride, mercury, nitrates, arsenic, uranium, and some organic compounds.

Figure 2-17 Small ion-exchange system installed for iron removal in Washington State

What are the key design requirements for ion exchange?

Important design criteria for ion-exchange systems focus on the volume of water that can pass through the system before the vessel must be regenerated. This volume of water is usually expressed as bed volume, with one bed volume being equal to the volume of ion-exchange resin in the vessel. Usually, several vessels are used for ion-exchange systems. The number of bed vessels between regenerations is critical to sizing of the resin beds and the number of vessels needed in the system.

Ion-exchange plants are often designed around the equipment selected. The designer often provides extensive water quality information, effluent treatment, and performance requirements to equipment suppliers when selecting the preferred equipment. Once the equipment is selected, the system design is completed.

Ion-exchange system performance is highly dependent on raw water quality and the target effluent concentration. For cation-exchange systems, the amount of calcium, magnesium, and other cations in the water as well as the pH are needed to estimate the run length for a resin bed. For anion-exchange systems, the pH, arsenic, nitrate, sulfate, chloride, fluoride, and other anion concentrations must be known.

In order to minimize waste, systems may collect portions of the backwash, regenerant, and rinse streams and reuse them in later regenerations. This requires an even greater understanding of the water quality to ensure long-term problems do not develop.

What types of residuals are associated with ion exchange?

Residuals from ion-exchange processes have very high levels of total dissolved solids (TDS) as well as high concentrations of the anions or cations being removed. Different resins require different concentrations of regenerant. Waste streams are typically 5 to 10% of water treatment plant production. However, many equipment suppliers recover backwash water, rinse water, and part of the regenerant stream to reduce the waste to less than 1% of production. In these minimized waste streams, TDS concentrations may approach or exceed 100,000 mg/L.

❖ How difficult is it to operate and maintain ion-exchange systems?

Most ion-exchange systems are automated to carefully control the regeneration operation. Major operational requirements include monitoring the raw and finished water concentrations and maintaining sufficient regenerant levels in the regeneration tanks. Maintenance is very important with ion-exchange systems. Because of the high concentrations of brine in the regenerant and waste streams, regenerant tanks, pumps, and piping must be periodically cleaned and flushed.

The amount of salt or other regenerant used is highly dependent on the quality of the water being treated; sulfate, alkalinity, and pH can dramatically affect anion-exchange regeneration frequency, while calcium and magnesium levels predominately affect cation exchange.

Iron and manganese can foul the resin, and silica adsorption onto resin surfaces has also been noted. If fouling occurs, the resin must be acid washed (cation resin), caustic washed (anion resin), or replaced to improve removal.

❖ How do commercially available ion-exchange systems differ?

Ion-exchange system design varies greatly from equipment supplier to equipment supplier, and many ingenious modifications have been made to improve efficiency, reduce regenerant, and minimize waste. Modifications include providing a packed bed system that does not require backwashing. Cocurrent (in the same direction as water flow) and countercurrent regenerating systems are available. Partial regeneration is sometimes used, and waste minimization strategies include multiple vessel systems that reuse brine.

Often, equipment is preselected, and the plant is designed around the selected equipment. A current list of vendors supplying ion-exchange systems can be found at http://sourcebook.awwa.org/.

MEMBRANE TREATMENT TECHNOLOGY

Treating groundwater using membranes is not uncommon for many applications, including softening, brine removal, TDS reduction, and removal of specific compounds such as iron and arsenic.

❖❖ How does membrane treatment work?

Figure 2-18 shows a simple process flow diagram for reverse osmosis (RO) or nanofiltration (NF) membrane treatment. Figure 2-19 shows a process flow diagram for ultrafiltration (UF) or microfiltration (MF) membrane treatment.

Membrane processes make use of semipermeable membrane material to physically filter suspended and, in some cases, dissolved compounds from water. Unlike filters in which filter media may be 500 to 1,000 times larger than the particles they are removing, membrane pores are smaller than the particles they retain on their surface.

Membranes are manufactured in a variety of configurations, materials, and pore size distributions. Membrane treatment selection for a particular drinking water application is based on a number of factors, including material(s) to be removed, source water quality characteristics, treated water quality requirements, membrane pore size, molecular weight cutoff (MWC), membrane materials, and system/treatment configuration.

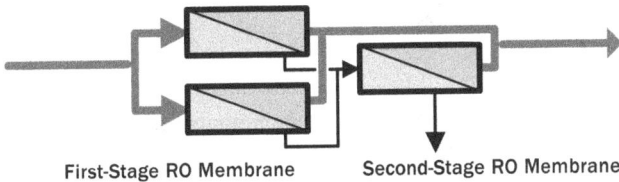

First-Stage RO Membrane   Second-Stage RO Membrane

Figure 2-18 RO or UF membrane process flow diagram. Shown as two stage system without pretreatment.

Courtesy of Paul Mueller, CH2M HILL

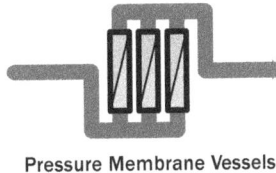

Pressure Membrane Vessels

Figure 2-19 MF or NF membrane process flow diagram. Shown without pretreatment.

Courtesy of Paul Mueller, CH2M HILL

❖❖❖ What types of treatment issues can membrane treatment effectively address?

Historically, the membrane technologies listed in the following paragraphs have been applied for specific drinking water uses (Figure 2-20). Typical membrane applications are listed in Table 2-5.

Reverse osmosis treatment in a high-pressure mode is used to remove dissolved metals and to remove salts from brackish water and seawater. Because of typical RO membrane pore sizes and size exclusion capability (in the metallic ion and aqueous salt range), RO filtration effectively removes almost all contaminants commonly found in water except volatile organic compounds.

Figure 2-20 Membrane technologies are commonly used to remove iron and arsenic from groundwater

Table 2-5   **Design Considerations for Membrane Treatment Systems**

Type	Reverse Osmosis	Nanofiltration	Ultrafiltration	Microfiltration
Typical applications	Desalination Brackish water High total dissolved solids Fluoride Metals Arsenic Synthetic organics	Softening Pesticides Nitrate Natural organics Color removal	Suspended solids Turbidity Viruses Bacteria Protozoan cysts Iron Manganese Arsenic Coagulated particles	Suspended solids Turbidity Viruses Bacteria Protozoan cysts Iron Manganese Arsenic Coagulated particles
Operating pressure range, psi	150–1200 (10–83 Bar) depending on total dissolved solids	50–150	15–40	3–40

Nanofiltration, also referred to as membrane softening or low-pressure RO, is used to remove calcium and magnesium ions (hardness), pesticides, nitrate in some waters, and natural organics. It is also used to control DBPs.

Ultrafiltration, characterized by a wide band of MWCs and pore sizes, is used to remove specific dissolved organics (e.g., humic substances, for DBP control in finished water) and to remove particulates.

Microfiltration, such as UF using low operating pressures, is used to remove particulates including pathogenic cysts.

❖ What are the key design requirements for membrane treatment?

Membrane treatment system design requires a thorough understanding of the operating requirements and the complexities of the particular system being used. As with ion-exchange plants, membrane plants are often designed around the specific equipment selected. The designer often provides extensive water quality information along with effluent treatment and performance requirements to equipment suppliers when selecting the equipment. Once the equipment is selected, the system design is completed.

Design considerations for membrane treatment systems include the following:

- Pretreatment requirements: These can vary from simple strainers to a full conventional treatment plant or complex chemical feed systems.
- Recovery rates and raw water feed: RO and NF systems may produce less than 50% of the raw water fed to the plant. Recovery rates range from 40 to 97%.
- Scaling and fouling: RO and NF systems require careful evaluation of scaling caused by manganese organic material and other contaminants. Fouling issues are also a major concern with UF and MF systems.
- Plant hydraulics: The range of operating pressures and transmembrane pressures varies considerably from system to system and during membrane operations. Hydraulics must be closely evaluated during the design phase.
- Water quality: Temperature is a key design criterion for all membranes, because recovery and flux rates are lower at colder temperatures. Critical design for membrane plants

often requires addressing cold weather conditions, rather than summertime peak-day conditions. Other water quality parameters are also very important.

- Equipment limitations: Some membrane systems have throughput and recovery limitations that must be adhered to for optimal performance.
- Backwash and cleaning intervals: These are important factors in providing reliable plant capacity and may require testing to develop accurate scaling and fouling rates for systems.

Pretreatment selection is a key component for all membrane systems. RO and NF systems require removal of suspended solids prior to the RO membrane in order to minimize fouling and ensure proper operation. Dissolved compounds that can precipitate, e.g., iron and manganese, are also usually removed before the RO or NF membranes. Scale-inhibiting chemicals are also added to protect membranes from plugging effects and fouling and/or scaling and to reduce O&M costs.

Pretreatment for UF and MF membrane systems is highly dependent on water quality. At a minimum, strainers are required to prevent harmful materials from entering the membrane system. Typically, this includes a 400- to 500-μm strainer. However, some operational evidence suggests that large numbers of diatoms may damage some membrane materials, and removal of particulate matter as small as 50 μm might be warranted. Typical operating pressures for each membrane type are given in Table 2-5.

❖❖❖ What types of residuals are associated with membrane treatment?

RO systems produce concentrated brine (typically three to five times the influent brine concentration), pretreatment residuals, and cleaning wastes, which include concentrated acid and chlorine wastes. UF systems include a residual stream with concentrations of calcium and magnesium that may be 3 to 10 times the raw water concentrations. UF residuals also include pretreatment and cleaning wastes similar to those found in RO systems. UF and MF system residuals may include concentrated turbidity, microorganisms, iron, manganese, and organic material. If coagulants are used in these systems, the coagulant will be present in the residual stream.

Membrane systems also use acid cleaning chemicals and concentrated chlorine to clean the membranes periodically. Because chlorine is routinely used as part of the membrane cleaning procedure, disposal of cleaning waste may include monitoring and limits on chlorinated organic compounds.

How difficult is it to operate and maintain membrane treatment systems?

RO systems require advanced pretreatment and have high capital costs and high O&M costs due, in large part, to energy and pumping requirements. Typically, operator labor is required for use of the pretreatment and cleaning chemicals and there is a greater focus on maintaining mechanical elements.

NF requires a relatively high degree of pretreatment, but required operator skills are fairly basic. O&M are required to optimize the pretreatment system, maintain chemical tanks, and maintain the plant's mechanical elements.

Operator requirements for UF and MF are fairly basic. However, if these systems are used for water with varying quality, extensive adjustment of chemical feeds and cleaning periods will be required.

How do commercially available membrane treatment systems differ?

Membrane equipment varies considerably from equipment supplier to equipment supplier. Systems may operate as pressure systems or, in the case of some UF and MF designs, under vacuum. The equipment is not interchangeable. In most cases, the membrane equipment is preselected, and the plant is designed around the selected equipment. A current list of vendors supplying membrane systems can be found at http://sourcebook.awwa.org/.

ELECTRODIALYSIS REVERSAL

How does electrodialysis reversal work?

Figure 2-21 shows a typical process flow diagram for electrodialysis reversal (EDR). EDR is essentially a membrane process, often using membranes with pore sizes similar to those in an RO membrane.

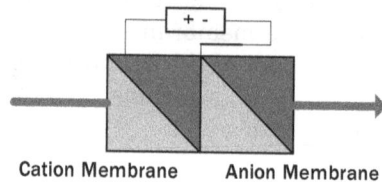

Figure 2-21 Electrodialysis reversal process flow diagram
Courtesy of Paul Mueller, CH2M HILL

Cation-selective and anion-selective membranes are paired within the EDR unit, which uses electrode polarity reversal to automatically clean membrane surfaces.

The electrodialysis process uses a driving force of direct current (DC) power to transfer ionic species from the feedwater through cation (positively charged ions) and anion (negatively charged ions) transfer membranes to a concentrate stream, creating a more dilute stream. Multiple stages are often used to achieve the desired effluent water quality.

The polarity of the DC power is reversed two to four times per hour. When the polarity is reversed, the dilute and concentrate compartments also are reversed. The alternating exposure of membrane surfaces to the dilute and concentrate streams provides a self-cleaning capability that enables purification and recovery of up to 94% of the feedwater.

❖ What types of treatment issues can electrodialysis reversal effectively address?

EDR systems can effectively remove most of the same contaminants that are removed by RO membrane systems. EDR is preferred over RO in desalination applications with high levels of silica. These systems also require similar types of extensive pretreatment and cleaning. EDR is most frequently used in desalination applications and to remove fluoride, radium, nitrate, arsenic, chloride, high TDS, and other inorganic compounds from groundwater.

What are the key design requirements for electrodialysis reversal?

Design criteria for EDR are similar to those for an RO system (described above), with a significant difference in the operating pressure. EDR systems typically operate at feed pressures less than 70 psig (5 bar) and do not require the same type of feed pump and materials to withstand the high pressures seen in RO systems.

Extensive pretreatment may be required prior to EDR. Suspended solids that are smaller than 10 μm must be removed, and fouling compounds such as iron and manganese are usually removed before EDR. In some difficult applications, pretreatment may include both a cation- and anion-exchange step. Pretreatment selection varies significantly depending on water quality and effluent requirements.

As with ion-exchange and membrane plants, EDR plants are often designed around the equipment selected. The designer often provides extensive water quality information as well as effluent treatment and performance requirements to equipment suppliers when selecting the equipment. Once equipment is selected, the system design is completed.

What types of residuals are associated with electrodialysis reversal?

A concentrated brine solution similar to RO residuals streams is produced. The pretreatment systems and cleaning systems also produce waste streams.

How difficult is it to operate and maintain electrodialysis reversal systems?

EDR systems require extensive maintenance and cleaning. Pretreatment systems may also require significant O&M.

How do commercially available electrodialysis reversal systems differ?

EDR equipment varies considerably from equipment supplier to equipment supplier. The equipment is not interchangeable. In most cases, EDR equipment is preselected and the plant is designed around the equipment. A current list of vendors supplying EDR systems can be found at http://sourcebook.awwa.org/.

SOFTENING PROCESSES

Water softening can be accomplished using cation-exchange systems, RO membranes, NF membranes, EDR systems, and excess lime softening, and pellet softening. Ion-exchange, RO, NF, and EDR systems have been discussed in previous sections. Consequently, this section focuses on lime and pellet softening.

❖❖ How do softening processes work?

Excess lime softening. Dissolved minerals and organic matter can be removed from water by adding excess lime or lime and soda ash. A lime softening process flow diagram is presented in Figure 2-22. In the lime–soda ash softening process, sufficient quantities of lime (calcium hydroxide) are added to the water to supersaturate the water with calcium and magnesium bicarbonates. As the pH rises above 10 to 11, the lime reacts with the bicarbonates, forming calcium carbonate and magnesium hydroxide precipitants. These precipitated compounds form sludge that can be removed in the water treatment process. The sludge is often retained in an upflow clarifier to help catalyze the precipitation of calcium and magnesium compounds and to clarify the water. After precipitating calcium and magnesium carbonate compounds, the water must be conditioned by lowering the pH and establishing a lower calcium carbonate precipitation potential.

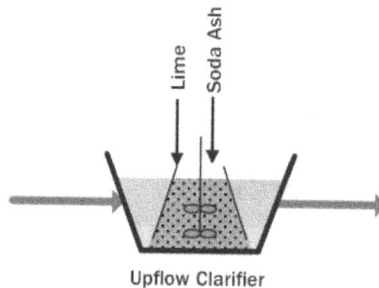

Lime   Soda Ash

Upflow Clarifier

Figure 2-22 Excess lime softening process flow diagram
Courtesy of Paul Mueller, CH2M HILL

Excess lime softening (which raises the pH to above 12) also provides microbial inactivation. With contact times of several hours at a high pH level, bacteria, viruses, and protozoan cysts can be inactivated and many heavy metals and organic compounds can be removed.

Pellet softening. Pellet softening operates on the same chemical principles as lime–soda ash softening but does not produce an undesirable sludge. Instead, the pellet softening system consists of a gravity or pressure tank in which calcium carbonate crystallizes on a suspended bed of fine sand and produces a gravel-sized pellet that can be removed.

First, the water is pretreated with caustic soda or lime to increase the pH for precipitation of calcium carbonate. The mixture is injected into the bottom of the reactor, and the flow moves quickly upward through a fluidized bed. The calcium carbonate precipitate forms on the sand grains to form pellets that are three to five times the size of the original sand media. Softened water requires lowering of the pH to establish a stable, lower calcium carbonate precipitation potential. As the pellets increase in size, they drop to the bottom of the fluidized bed where they are removed and replaced with fresh sand media.

This treatment method is generally only successful at removing calcium bicarbonate hardness. It is not appropriate for systems with high magnesium content, because of potential magnesium hydroxide fouling of the reactor. Iron removal can take place concurrently with the softening, while manganese removal usually requires posttreatment. Postfiltration may be required with pellet softener systems, because the relatively short contact times often make it difficult to stabilize the water (so that it does not continue to precipitate) after softening.

❖ What types of treatment issues can softening processes address?

Hardness is comprised of divalent cations that usually include calcium carbonate and magnesium carbonate compounds. Iron and manganese in their reduced forms are also divalent cations and can be removed in softening. Arsenic can be coprecipitated with lime softening as well. Pellet softening systems are typically limited to reduction of calcium carbonate and iron removal. Fouling of pellet softeners is common when the pH is raised to precipitate manganese or magnesium carbonate compounds.

❖ What are the key design requirements for softening processes?

Water chemistry, overflow rates, weir loading, solids retention, chemical feed design, and residuals handling systems are all important considerations with lime softening systems. Lime and soda ash systems must be designed with feed capacities that will achieve the desired pH for precipitation in the softening system. Soda ash softening is a dry feed system that requires dry chemical feed storage and feeding equipment, mixing, and conveyance to the softener. Lime may be fed as pebbles of hydrated lime. This pebbled lime requires dry pebble storage, a dry feed system, and a slaker, which hydrates the lime into a slurry. The hydrated lime slurry is then conveyed to the softening unit. Both systems require frequent maintenance and cleaning to prevent clogging.

The upflow clarifier must be designed to retain sludge in a fluidized bed without washing out the sludge at the top of the clarifier. Weir overflow rates must be low enough to prevent precipitate carryover. Also, sludge removal must be designed to allow frequent withdrawal from the clarifier without upsetting the fluidized sludge bed. Posttreatment is often accomplished with carbon dioxide. Feed systems need to be designed to allow sufficient gas transfer to lower the pH to the desired posttreatment level.

Pellet system design criteria include flow rates through the softener unit that will maintain a fluidized bed with fresh sand media and softened pellets without washing them out of the top of the tank. Removal of pellets from the base of the unit must be addressed during the design phase to ensure the fluidized bed is not disrupted. Design of the chemical feed system to raise pH to the desired level and for posttreatment is also important. Postfiltration must be designed to retain precipitated calcium carbonate particles that escape the unit.

❖ What types of residuals are associated with softening processes?

Residuals include precipitates of calcium and magnesium carbonate as well as lower concentrations of other precipitated compounds. Lime softening systems produce a liquid sludge that is often dried on site or trucked to an off-site drying location. Sludge can also be directly applied to land surfaces and incorporated as a soil amendment. Pellet softeners produce a calcium carbonate pellet that can be drained of freestanding water and readily reused as a soil amendment. For many softening systems, the residuals are readily used in area agricultural applications.

❖ How difficult is it to operate and maintain softening systems?

The chemical feed systems may include a lime slaker, which requires careful operation and oversight as well as routine maintenance. Hydrated lime and soda ash readily precipitate in pipes and on clarifier walls, and frequent cleaning is required. Dry chemical feed systems must be cleaned frequently and require dust control.

Solids contact clarifiers must be carefully monitored to ensure solids do not pass over the collection weirs. Flow rates through the clarifier or softener must be monitored frequently as does the sludge blanket level in an upflow clarifier.

The pH of the water in the softening system must be carefully monitored and controlled in order to achieve the proper amount of precipitation in the softening unit. Calcium carbonate precipitation potential and posttreatment pH must be carefully monitored.

Residuals handling requires routine maintenance to control buildup of the precipitated materials. Sludge handling systems must be routinely maintained because of the high concentrations of solids in softening sludge.

❖ How do commercially available softening systems differ?

Suppliers of softening systems include those supplying ion-exchange, membrane, and EDR equipment as well as lime and/or soda ash equipment. Systems vary depending on the softening method used and among individual suppliers. Smaller systems may be provided with a complete unit. Larger systems, especially those for lime and/or soda ash softening, often require a clarifier design for which the supplier provides the mechanical and control equipment as well as the chemical feed systems. A current list of vendors supplying softening systems can be found at http://sourcebook.awwa.org/.

AERATION AND DEGASSING TREATMENT TECHNOLOGY

Aeration may be used to remove offensive tastes and odors that result when gases from decomposing organic matter are dissolved;

to reduce or remove objectionable amounts of carbon dioxide, hydrogen sulfide, and similar products; and to introduce oxygen to assist in iron and/or manganese removal.

❖❖ How does aeration and degassing treatment work?

Figure 2-23 provides a process flow diagram for packed tower aeration. Aeration works by transferring gas from the water into the air or from the air into the water. Aeration is accomplished by spraying water, bubbling or injecting air into the water stream, or cascading water over trays or a loose media that breaks up the water flow into smaller drops, thus creating a larger water-to-air surface. Aeration can also be accomplished using mechanical aerators. With all aeration system, the water is exposed to atmospheric pressure and the water is collected and repumped to meet system pressure requirements.

Packed tower aeration (PTA) involves passing water down through a column of packing material while forcing air up through the packing media. Forced-air systems can also be used with tray aerators.

❖❖ What types of treatment issues can aeration and degassing treatment effectively address?

Aeration is used to remove hydrogen sulfide, volatile organic chemicals, trihalomethanes, carbon dioxide, and radon. Aeration can also be used to add oxygen to water for iron oxidation or to allow bacterial growth to occur in a biological filtration system.

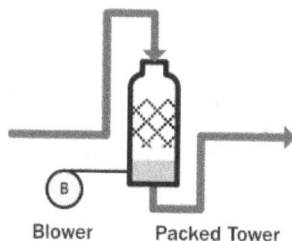

Blower   Packed Tower

Figure 2-23 Aeration process flow diagram
Courtesy of Paul Mueller, CH2M HILL

❖❖❖ What are the key design requirements for aeration and degassing treatment?

Generally, aeration is feasible for removing compounds with a Henry's constant greater than 100 (expressed in atm mol/mol) but not normally feasible for removing compounds with a Henry's constant less than 10. For values between 10 and 100, PTA or other forced-air systems are used and should be evaluated using pilot studies. Table 2-6 shows categories of aerated compounds in water that are commonly aerated.

Aeration system design depends on the type of aeration system considered, but design procedures are generally well established for different aeration system and explained clearly in several texts. Because the Henry's constant for each compound decreases with temperature, the system should be designed for the coldest air and water temperatures expected during operation.

The pH of the water is critical for removal of carbon dioxide and hydrogen sulfide. As pH increases, carbon dioxide in water becomes less dissociated as carbonic acid (H_2CO_3) and hydrogen sulfide is

Table 2-6   Aerated Compounds in Water (based on Henry's constant at 20°C)

Henry's Constant	Compounds
>100; readily aerated	Vinyl chloride, oxygen, nitrogen, methane, ozone, toxaphene, carbon dioxide, radon, carbon tetrachloride, tetrachloroethylene, trichloroethylene, hydrogen sulfide, chloromethane, 1,1,1-trichloroethylene, toluene, 1,2,4-trimethylbenzene, benzene, 1,4-dichlorobenzene, chloroform
>10 and <100; potentially removed with forced-air aeration	1-2-dichloromethane, 1,1,2-trichloroethene, sulfide dioxide, bromoform
<10; not readily aerated	Ammonia, pentachlorphenol, dieldrin, benzene, aldicarb, chlordane, polychlorinated biphenols

Table 2-7   Percent of Carbon Dioxide and Hydrogen Sulfide Available for Aeration Removal at Various pH Levels

Compound	pH 6	pH 6.5	pH 7	pH 7.5	pH 8	pH 8.5
Carbon dioxide, % as H_2CO_3	75	50	20	10	5	0
Sulfide, % as hydrogen sulfide	80	60	30	15	7	3

Source: GE Handbook of Industrial Water Treatment (1997–2009)

converted to sulfide. The percent available for removal by aeration for both of these compounds that can be achieved at different pHs is shown in Table 2-7.

The tower air outlet in PTA systems must be designed to prevent noise from becoming a nuisance to neighbors. Sound regulations vary by site and zoning, but often the design must meet a decibel level at the property line. Air quality permitting may also be required, depending on the contaminants removed and the location of the facility.

❖❖ What types of residuals are associated with aeration and degassing treatment?

Residuals from aeration systems are off-gases. In some applications, the off-gas must be collected and contaminants removed prior to discharge to the atmosphere. GAC canisters are most commonly used to collect and treat off-gases. Specific requirements for off-gas treatment from aeration facilities vary by location and with air quality standards.

❖❖ How difficult is it to operate and maintain aeration and degassing systems?

Aeration systems are easy to operate. Periodic cleaning of the aeration media with a dilute acid may be required. Maintenance of blowers, pumps, and mechanical components is required.

❖❖ How do commercially available aeration and degassing systems differ?

There are numerous aeration equipment suppliers. Systems vary by type and among suppliers. Designs for the distribution of water and media types used to achieve gas transfer are often proprietary. A current list of vendors supplying aeration systems can be found at http://sourcebook.awwa.org/.

DISINFECTION, OXIDATION, AND CORROSION CONTROL: CHEMICAL TREATMENT TECHNOLOGIES

Various chemical feed systems including disinfection, oxidation, and corrosion control feed systems are discussed in this section. Table 2-8 provides a summary of applications, forms available, and design and operational issues for various chemical feed systems used in groundwater treatment applications (Figure 2-24). Because the number and availability of equipment suppliers are so large, equipment supplier information for chemical feed systems is not provided here. A current list of vendors supplying specific chemical feed systems can be found at http://sourcebook.awwa.org/.

Figure 2-24 Chemical feed systems require careful selection of materials, process needs, and safety considerations

Table 2-8   Chemical Feed System Comparison

Chemical Type	Forms Available	Application	Design Issues	Operational Issues
Permanganate	Potassium permanganate (mixed in batches or dissolved in a saturator) Sodium permanganate (20% or 40% solution)	Iron and manganese removal Color removal Coagulant aid (must be fed prior to filter system)	Materials must be chemically compatible Containment needed Care must be taken to prevent overfeeding or siphoning	Pink or purple water can result from overfeeding Health hazards can result from overfeeding Strong oxidant that is a safety hazard Stains easily Stains can be removed with ascorbic acid Must use appropriate PPE
Chlorine	Sodium hypochlorite (purchased or generated on site) available in 0.8, 5, 6, 12.5, and 15% concentrations Calcium hypochlorite available as tablet or powder, 65 to 75% concentrations Chlorine gas (fed as gas under vacuum or evaporated as a liquid)	Primary disinfectant for microbial contaminants Residual disinfection Hydrogen sulfide, iron, manganese oxidation Pre-oxidant, color removal, coagulant aid	Materials must be chemically compatible Containment needed, except with 0.8% on-site generation Care must be taken to prevent overfeeding or siphoning Disinfection or oxidation by-products will form	Strong oxidant that is a safety hazard Must use appropriate PPE

Table 2-8 Chemical Feed System Comparison

Chemical Type	Forms Available	Application	Design Issues	Operational Issues
Chloramine	All chlorine combinations listed previously in combination with ammonia feed	Residual disinfection	Materials must be chemically compatible	Nitrification may occur in the distribution system
	Ammonia may be used as a gas (anhydrous) or a liquid; liquid or aqueous ammonia is typically available in concentrations ranging from 20 to 50%	Reduction of chlorinous tastes	The ratio of chlorine to ammonia applied is critical to stabilizing the chloramine compounds	Rubber will degrade in the presence of chloramine
		Reduction of disinfection by-products	Care must be taken to prevent overfeeding or siphoning	Aquariums and dialysis units must remove ammonia prior to use; people and fish have been killed when systems have switched to chloramines without effective public notice and education
			Disinfection or oxidation by-products will form	Chloramine may form di-chloramine or tri-chloramine and may result in strong chlorinous tastes and odors
				Strong oxidant that is a safety hazard
				Must use appropriate PPE
Chlorine dioxide	Generated on site with sodium chlorate and acid	Oxidation of organic compounds, hydrogen sulfide, iron, and manganese	Materials must be chemically compatible	Strong oxidant that is a safety hazard
	Generated on site with sodium chlorite and chlorine	Color removal	Containment needed	Must use appropriate PPE
	Available in tablet form and mixed in batches	Reduction of disinfection by-products	Care must be taken to prevent overfeeding or siphoning	
		Primary disinfectant for microbial contaminants	Limited to 1 mg/L maximum residual for chlorine dioxide, chlorite, and chlorate concentrations	
			Disinfection or oxidation by-products will form	

(continued)

Table 2-8   Chemical Feed System Comparison (continued)

Chemical Type	Forms Available	Application	Design Issues	Operational Issues
Ozone	Generated on site with oxygen and high-voltage electricity Systems include oxygen feed system, ozone generator (10 to 15% ozone concentration by weight), injection system, contactor, off-gas collection and destruct systems, and monitoring equipment	Oxidation of organic compounds, hydrogen sulfide, iron, and manganese Taste and odor reduction Color removal Reduction of disinfection by-products Primary disinfectant for microbial contaminants	Materials must be chemically compatible Injectors and contactors must be designed for effective ozone gas transfer Moisture must be kept out of ozone generation equipment Liquid oxygen most often used for larger systems and must be evaporated Care must be taken to prevent overfeeding Manganese or manganese-containing media can be oxidized to permanganate if overdosed Disinfection or oxidation by-products will form	Strong oxidant that is a safety hazard Air exposure to ozone will cause health hazard Must use appropriate PPE
Ultraviolet light	Low pressure, high output Medium pressure	Disinfectant for bacteria and cysts High doses or use in combination with peroxide can oxidize many organic compounds	Not effective for virus inactivation Dosing validation for microbial contaminants varies considerably by state	Bulb fouling and maintenance issues may drive design and selection of reactor type Electrical hazard exists for routine maintenance operations Calibration and sensor checks required to demonstrate disinfection dose

Table 2-8   Chemical Feed System Comparison

Chemical Type	Forms Available	Application	Design Issues	Operational Issues
Caustic	Liquid concentrations ranging from 30 to 50%	pH adjustment Precipitation Regeneration of some ion-exchange systems	Materials must be chemically compatible Containment needed Solution must be kept warm to prevent precipitation Care must be taken to prevent overfeeding or siphoning	Strong reactive substance that is a safety hazard Must use appropriate PPE
Lime	Pebble (must be slaked prior to use) Quicklime or calcium oxide solids or liquid (only available in some markets)	pH and alkalinity adjustment Precipitation	Materials must be chemically compatible Feed lines will encrust with precipitates and must be flushed or flexible to maintain flow Containment needed Care must be taken to prevent overfeeding or siphoning	Strong reactive substance that is a safety hazard Slakers require constant oversight and monitoring to prevent encrustation Feed lines and equipment require frequent routine maintenance to prevent build-up of calcium oxide Must use appropriate PPE

(continued)

Table 2-8 Chemical Feed System Comparison (continued)

Chemical Type	Forms Available	Application	Design Issues	Operational Issues
Soda ash	Powder that is mixed with water and fed as a liquid	pH and alkalinity adjustment Precipitation	Solubility of soda ash varies considerable from about 8% at 5°C to about 23% at 25°C Dry chemical feed system design and control needed for large systems Feed lines will encrust with precipitates and must be flushed or flexible to maintain flow Care must be taken to prevent overfeeding or siphoning	Alkaline substance that is a safety hazard Dry feed systems require oversight and monitoring to prevent encrustation Feed lines and equipment require frequent routine maintenance to prevent build-up of carbonate Must use appropriate PPE
Carbon dioxide	Gas (will form carbonic acid in water)	pH and alkalinity adjustment	Materials must be chemically compatible Injectors and contactors must be designed for effective gas transfer Care must be taken to prevent overfeeding Disinfection or oxidation by-products will form	Air exposure to excess carbon dioxide will cause health hazard Must use appropriate PPE
Calcite (limestone)	Pebble	pH and alkalinity adjustment	Contactors require strainers before and after the contactor Contactor depth and contact time vary with water quality Effective for low pH, low to moderate alkalinity, and low to moderate hardness waters	Must monitor the calcite level and replace to maintain minimum media level

Table 2-8   Chemical Feed System Comparison

Chemical Type	Forms Available	Application	Design Issues	Operational Issues
Acids	Sulfuric acid (concentrations vary) Muratic acid (concentrations vary) Citric acid (concentrations vary) Ascorbic acid (concentrations vary)	pH adjustment Cleaning of resins and media Regeneration of some ion-exchange systems	Materials must be chemically compatible Containment needed Care must be taken to prevent overfeeding or siphoning	Strong reactive substance that is a safety hazard Must use appropriate PPE
Phosphates	Orthophosphates (phosphoric acid): liquids and powders Polyphosphates: liquids and powders Phosphate blends: liquids and powders Hexametaphosphates: liquids and powders	Corrosion control (orthophosphate) Iron and manganese sequesterant (polyphosphate) Prevention of scale formation (hexametaphosphate)	Appropriate feed system design for application type is needed Care must be taken to prevent overfeeding or siphoning Materials must be chemically compatible Containment needed for some products	Routine maintenance Some products may pose safety hazard Must use appropriate PPE
Silicates	Sodium silicate (pH and concentrations vary)	Corrosion control Sequestering of iron and manganese	Materials must be chemically compatible Solution must be kept warm to prevent precipitation Solution is very viscous Containment needed Care must be taken to prevent overfeeding or siphoning	Strong reactive substance that is a safety hazard Must use appropriate PPE Routine maintenance required to clean feed lines and equipment

PPE—personal protective equipment

REFERENCES

AWWA (American Water Works Association). 1999. *Water Quality and Treatment*, 5th ed. McGraw-Hill: New York.

AWWA. 1999. *Water Treatment Plant Design*, 5th ed. McGraw-Hill: New York.

Gage, B., et al. 2001. Biological Iron and Manganese Removal, Pilot and Full Scale Applications. Ontario Water Works Association Conference, May 3, 2001.

General Electric. 1997–2009. *Handbook of Industrial Water Treatment*. General Electric Company: Fairfield, Conn. www.gewater.com/handbook/index.jsp. Accessed: July 16, 2009.

Great Lakes–Upper Mississippi River Board of State and Provincial Public Health and Environmental Managers. 2003. Recommended Standards for Water Works. Health Research Inc., Health Education Services Division: Albany, N.Y.

Knocke, W.R. 1990. *Removal of Soluble Manganese From Water by Oxide-coated Filter Media*. AWWA Research Foundation and AWWA: Denver, Colo.

Knocke, W.R., S.C. Occiano, and R. Hungate. 1991. Removal of Soluble Manganese by Oxide-Coated Filter Media: Sorption Rate and Removal Mechanism Issues. *Jour. AWWA*, 83:8:64.

CHAPTER THREE

Disinfection of Groundwater

Groundwater is disinfected to achieve three objectives:
- Systems that are not vulnerable to contamination may provide disinfection for general practice and to maintain a distribution system residual.
- Systems that are vulnerable to contamination under the Groundwater Rule, that have fecal coliform in their source water, or that have had positive results indicating the presence of coliform in their distribution system practice disinfection for virus inactivation and distribution system residual.
- Systems that are classified as under the influence of surface water must meet surface water disinfection criteria and provide a disinfection residual.

The choice of primary and secondary disinfectant varies among these three treatment objectives. For example, ultraviolet (UV) light disinfection is particularly effective for *Cryptosporidium* and *Giardia* inactivation but not for viruses. Therefore, UV is used by groundwater systems only if they are under the influence of surface water.

TREATMENT ALTERNATIVES FOR DISINFECTION

Table 3-1 summarizes treatment alternatives for groundwater systems that disinfect. It should be noted that some groundwater systems do not disinfect their water supplies.

Chlorine, which is the most commonly used disinfectant, is the only disinfectant that can be used as both a primary and secondary disinfectant for all groundwater disinfection objectives. However, chlorine alone is not always adequate, and for groundwater systems classified as under the influence of surface water, additional treatment is often required.

The choice of primary and secondary disinfectants is largely based on water quality parameters and treatment objectives. Each disinfectant and their applications are discussed in this chapter.

Table 3-1   Disinfectant Alternatives for Groundwater Systems

Treatment objective	General Disinfection and Residual Disinfection	Disinfection of Viruses and Residual Disinfection	Disinfection for Systems Under the Influence of Surface Water
Primary disinfectant	Chlorine Ozone Chlorine dioxide	Chlorine Ozone UV	Chlorine* Ozone Chlorine dioxide UV
Secondary (residual) disinfectant	Chlorine Chloramine	Chlorine Chloramine	Chlorine Chloramine

*Must provide additional disinfectant for *Cryptosporidium* inactivation if unfiltered or if *Cryptosporidium* levels in raw water require it.

CHLORINE

Chlorine can be applied in a number of ways: as chlorine liquid or gas, as sodium hypochlorite, or as calcium hypochlorite. In addition, sodium hypochlorite can be purchased as a solution or generated on site. Once applied to water, chlorine forms hypochlorous acid and hypochlorite ions, regardless of the form applied. If ammonia is present in the water, enough chlorine must be added to fully react with the ammonia before free chlorine is formed. Hypochlorous acid and hypochlorite are forms of free chlorine. Ammonia reactions with chlorine are described in the chloramine section of this chapter.

Chlorine has a maximum residual disinfectant level of 4 mg/L and will form chlorinated disinfection by-products (DBPs), including trihalomethanes and haloacetic acids, which have regulated maximum contaminant levels (MCLs) (see Chapter 1).

Chlorine Gas Systems

Chlorine gas systems generally include 150-lb (68-kg) or 2,000-lb (909-kg) cylinders. A vacuum line with an automatic shut-off valve should be directly affixed to the gas cylinder. This valve will close if the system loses vacuum, preventing a large leak from the cylinder. In the event of a fire and to prevent the gas cylinder from exploding, gas chlorine cylinders include a fusible plug that will melt and release gas (Figure 3-1).

A flow-control valve and meter are used to control the amount of gas entering the water. The vacuum is formed by a venturi meter, which is usually located on a feedwater line. The venturi mixes the

Figure 3-1 Gas chlorination system in 2005, now replaced with on-site hypochlorite generation system at Batavia, Illinois

gas chlorine with the feedwater, and the chlorine solution is injected into the water supply. To accomplish this, the feedwater line must have a higher pressure than the main water supply at the point of injection.

Chlorine gas is colorless, except at high concentrations when it appears green. It is also heavier than air and tends to accumulate in low or poorly ventilated spaces. Chlorine gas is a strong oxidizer and may react with flammable materials.

Chlorine is a toxic gas that irritates the respiratory system. Coughing and vomiting may occur at levels of 30 ppm and lung damage may occur at 60 ppm; exposure can be fatal at concentrations of approximately 1,000 ppm. Breathing lower concentrations can aggravate the respiratory system, and exposure to the gas can irritate the eyes. If exposed to chlorine gas, it may burn the smell receptors and, as a result, exposed people may not be able to detect it. Because chlorine gas forms a white cloud when it reacts with ammonia, ammonia vapor is often used by to detect leaks.

A chlorine gas sensor and alarm should be part of any chlorine gas installation. Personal protective equipment including a respirator should be provided to personnel accessing areas where chlorine gas is stored and used.

Many regulations and fire codes apply to gas chlorine installations. Local fire and building officials should be consulted prior to installing gas chlorine systems to ensure compliance. Chlorine gas installations may be subject to containment, public disclosure, hazardous material, and other requirements.

Recommended materials for use in chlorine gas systems are included on the chemical compatibility chart in Appendix A.

Sodium Hypochlorite Systems

Sodium hypochlorite is a liquid form of chlorine. It is a strong oxidizer, and products of the oxidation reactions are corrosive. Solutions can burn skin and cause eye damage, particularly when used in concentrated forms. However, as recognized by the National Fire Protection Association (NFPA, 2010), only solutions containing more than 40% sodium hypochlorite by weight are considered hazardous oxidizers. Solutions containing less than 40% sodium hypochlorite are classified as moderate oxidizing hazards (NFPA, 2010). Sodium hypochlorite solutions are typically stabilized using a caustic soda.

Sodium hypochlorite is purchased in concentrations ranging from 5 to 15%. It can also be generated on site, typically in concentrations of 0.7 to 0.8%; it can also be generated at 12 to 15% concentrations.

Sodium hypochlorite is often fed using a chemical feed pump system that injects the solution into the water supply (Figure 3-2). A 30-day supply of solution is usually provided, although the solution strength will degrade over time. The solution tank should be constructed of a chemically compatible material. Concentrations greater

Figure 3-2 Sodium hypochlorite tank (on right) and fluoride saturator in Battle Ground, Washington

than 1% strength require secondary containment. The chemical feed pump system should include a pump, suction and discharge piping, pump controls, backpressure valve, and a secondary method to prevent siphoning of the solution into the water supply.

A method for measuring the amount of chlorine entering the system should be provided and may include a flowmeter, scale for the solution tank, or liquid level metering system. Flowmeters must be compatible with the chlorine solution system, otherwise a system that does not come in contact with the fluid can be used. Because sodium hypochlorite is subject to off-gassing, pumps and chemical feed systems should incorporate measures to reprime pumps and minimize impacts from off-gassing.

On-site generation of sodium hypochlorite solutions has increased in popularity for water treatment disinfection and residual applications (Figures 3-3 and 3-4). Generation systems use low-voltage electricity to produce chlorine from a dilute brine solution. Systems that produce concentrations of 0.7 to 0.8% include a brine tank, dilution system, power conditioning system, control system, and generation cell. The generation cell includes an anode and cathode that will form the hypochlorite solution when the electricity is supplied. The solution is put into a tank and fed with a chemical feed system,

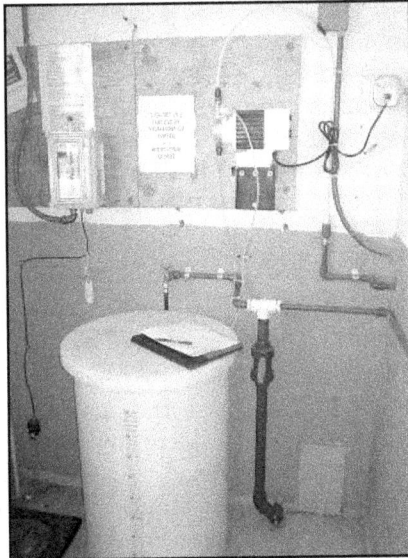

Figure 3-3 Sodium hypochlorite tank and feed pump

Figure 3-4 On-site sodium hypochlorite generator in Lakewood, Washington

which is similar to other sodium hypochlorite feed systems. Hydrogen gas is generated as a by-product of these systems and requires venting. Systems are available with production rates of 6 lb (2 kg) per day to several hundred pounds per day and more for municipal applications.

On-site generation of stronger concentrations of sodium hypochlorite is generally reserved for applications that require more than 100 lb/day (50 kg/day) and is commonly used in larger systems. These stronger on-site generation systems produce sodium hypochlorite with 10 to 15% concentrations. In addition to brine, systems also include caustic soda and acid feed systems and require containment for solutions.

Calcium Hypochlorite

Calcium hypochlorite, a solid form of chlorine, is available in tablet and powder forms. Calcium hypochlorite concentrations are typically 65% chlorine. Disinfection systems that use calcium hypochlorite often use the tablet form. The tablets are eroded by a controlled flow of water, forming a concentrated chlorine solution that is then pumped into the water using a chemical feed system, similar to those used with sodium hypochlorite. Solution tanks require secondary containment, and materials must be compatible with the solution (Figure 3-5).

Figure 3-5 A simple venting system keeps chlorine fumes from corroding well
house equipment

OZONE

Ozone is often used to disinfect groundwater when there is a specific
treatment objective such as removal of iron, manganese, taste, odor,
color, or high levels of natural organic matter (NOM) or when ozonation is part of a biological treatment process. Ozone is a strong oxidizing gas that must be generated on site; it is an effective disinfectant
for bacteria, viruses, and parasites regulated in water supplies.

Ozone is formed by applying a high-voltage arc of electricity to
oxygen molecules (O_2). These molecules cleave and reform partially
as ozone (O_3). An ozone system includes an oxygen supply, high-voltage generator, ozone injection system, off-gas destruction unit, ozone
gas and water analyzers, and control system.

The oxygen supply can be compressed oxygen gas, liquid oxygen,
or air. Systems that use air must dry and cool the air to remove mois-
ture and, in turn, prevent damage to the generator. Liquid oxygen
systems include a vaporizer and heater. Nitrogen is often added to
the oxygen stream to increase the percent of ozone formed in the generator. Ozone concentrations range from a few percent to nearly 20%,
but typical municipal systems produce approximately 12%.

The arrangement and design of the high-voltage electrodes in the generator vary by manufacturer's design. For cooling, a gas–water tube heat-exchanger is usually required. Typical gas pressures are less than 30 psi (2 bar) for oxygen systems and less than 45 psi (3 bar) for air systems. Power requirements may be significant when generating ozone, making it necessary to install several megawatts of electrical power in large facilities. Power is normally applied to the generator as single-phase AC current at 50 to 8,000 Hz and peak voltages between 3,000 and 20,000 V.

Ozone can be injected by diffusing bubbles through a column of water or by injecting a solution of ozone containing water into a main water supply, similar to the application of chlorine gas. Both methods can achieve greater than 95% ozone gas transfer.

Because all of the ozone will not transfer to the water, the gas must be collected from the water and destroyed. The ozone destruct system may include heat and a catalyst such as manganese dioxide. In addition, the system often includes a blower to create a vacuum and minimize the potential for ozone leaks to the atmosphere.

Because of its strong oxidizing properties, ozone is a primary irritant that affects the eyes and respiratory system and can be hazardous at even low concentrations. The Occupational Safety Health Administration (OSHA) has established a permissible exposure limit (PEL) of 0.1 ppm (OSHA, 2009, Table Z-1), calculated as an 8-hr time-weighted average. Higher concentrations are especially hazardous, and the National Institute for Occupational Safety and Health has established an immediately dangerous to life and health limit (IDLH) of 5 ppm. The work environments where ozone is used or where it is likely to be produced should have adequate ventilation and an ozone monitor that will alarm if the concentration exceeds the OSHA PEL. Continuous monitors for ozone are available from several suppliers.

CHLORINE DIOXIDE

Similar to ozone, chlorine dioxide is usually applied where there are additional treatment needs. In addition to disinfection, chlorine dioxide is also used to

- oxidize iron and manganese (especially when high levels of NOM is present);
- oxidize hydrogen sulfide;
- control tastes, odors, and colors; and
- oxidize phenolic compounds.

Chlorine dioxide is generated on site because at a concentration of 15% at standard temperature and pressure it is explosive under pressure and is not shipped. High-purity chlorine dioxide gas (7.7% in air or nitrogen) can be produced by reacting chlorine gas with solid sodium chlorite. Methods include mixing sodium chlorite with one of the following: hydrochloric acid, sodium hypochlorite, or chlorine gas. Chlorine dioxide can also be produced by reducing sodium chlorate in a strong acid solution using a suitable reducing agent such as hydrochloric acid or sulfur dioxide. A second method uses sodium chlorate, hydrogen peroxide, or sulfuric acid. Most methods use sodium chlorite combined with an acid or with sodium hypochlorite. Chlorine dioxide can also be produced by the electrolysis of a chlorite solution.

Design and operation requirements for chlorine dioxide systems vary significantly by type of system. Chlorite, a by-product of chlorine dioxide production, has an MCL of 1 mg/L, which may limit the dose of chlorine dioxide that can be applied to water. The amount of chlorite produced in the generation of chlorine dioxide varies by the method used to produce it.

Hydrochloric acid systems are generally smaller, generating less than 25 lb/day (12 kg/day). Systems using sodium hypochlorite have capacities of up to 1,000 lb/day (500 kg/day), and gaseous chlorine systems have capacities of up to several thousand pounds (kilograms) per day. Some systems incorporate more than one process to produce chlorine dioxide, although these are generally larger systems.

UV LIGHT

Ultraviolet (UV) light is electromagnetic radiation with a wavelength shorter than that of visible light, i.e., in the range of 10 to 400 nm. The electromagnetic spectrum of UV light can be subdivided in a number of ways. The draft International Standards Organization (ISO, 2007) standard on determining solar irradiances describes the ranges shown in Table 3-2.

Water can be disinfected using light with UVC wavelengths from 200 to 280 nm and, to a lesser extent, UVB wavelengths from 280 to 300 nm. UV light in the germicidal wavelengths, which is generated in lamps filled with mercury vapor, damages nucleic acid, preventing microbes from replicating. In addition to wavelength, the effectiveness of UV light as a disinfectant is determined by the dose. The UV dose is the product of UV intensity times the time of exposure and is often expressed in milli-Joules per centimeter squared (mJ/cm^2).

Table 3-2   Types of UV Light and Their Ranges

Type of UV Light	Abbreviation	Wavelength Range, nm
Ultraviolet A, long wave, or black light	UVA	400–320
Near	NUV	400–300
Ultraviolet B or medium wave	UVB	320–280
Middle	MUV	300–200
Ultraviolet C, short wave, or germicidal	UVC	280–100
Far	FUV	200–122
Vacuum	VUV	200–10
Extreme	EUV	121–10

UV disinfection is not particularly effective for virus inactivation. UV is also used in some applications to quench ozone or chloramines in water.

Disinfection is also affected by the UV transmittance of the water being treated, as influenced by the presence of UV light–absorbing organic chemicals. In addition, deposits that build up on the sleeves protecting the lamps also reduce the amount of UV light reaching the target pathogens in the water flow. Calcium, magnesium, and iron contribute to sleeve fouling.

The process equipment needed for UV disinfection includes the UV lamps; quartz sleeves protecting each lamp; UV sensors to measure lamp output; a cleaning system for the sleeves; a reactor vessel that houses the lamps, sleeves, and wiping equipment; a power supply system; lamp ballasts; UV transmittance monitor; and instrumentation and controls. There are two main types of UV lamps used for disinfection: low pressure and medium pressure. Low-pressure lamps produce a single peak of UV output at 254 nm; medium-pressure lamps produce a broader spread of UV output, between 210 and 310 nm.

UV intensity gradually decreases as the lamps are operated; UV output decreases as lamps age, reducing the UV dose for a given set of conditions (e.g., UV transmittance, flow rate, and lamp settings). Lamps (or more precisely their sleeves), which must be routinely cleaned, are encased in quartz sleeves and are not in direct contact with the water. Lamps are replaced when they are no longer able to produce the design level of UV intensity.

The design of UV systems should incorporate the expected UV transmittance of the water as well as allow for fouling and be operational within the expected flow and temperature ranges. Regulatory requirements for UV systems vary by state. Start-up and shut-down operations are critical for UV disinfection, because the bulbs take some time to warm up and stop disinfecting immediately when turned off.

CHLORAMINE

Chloramine is a combination of chlorine and ammonia. The term *chloramine* encompasses three chloramine compounds: monochloramine (NH_2Cl), dichloramine ($NHCl_2$), and trichloramine (NCl_3). Monochloramine is the desired disinfectant for systems that use chloramines, because dichloramine and trichloramine produce chlorinous tastes and odors. In addition, trichloramine can cause a burning sensation to the eyes, especially during showering. Monochloramine generally has a milder chlorine smell compared to free chlorine and lasts longer than free chlorine in the distribution system, but it is a weaker oxidant.

Chloramine is usually applied to reduce specific DBPs—trihalomethanes and haloacetic acids—which have MCLs. Chloramines produce a variety of nonregulated DBPs and can degrade rubber and elastomeric compounds. Chloramines can also significantly interfere with aquaculture and kidney dialysis if the chloramine compounds are not removed. Public notification and careful planning are needed before using chloramine compounds.

Chloramines, which are normally formed by adding a specific amount of ammonia to prechlorinated water, can also be formed by chlorinating water that contains naturally occurring ammonia. By combining chlorine with ammonia in water at specific ratios, the type of chloramine compounds can generally be controlled. At a ratio of 5:1 (milligrams per liter of residual free chlorine to milligrams per liter ammonia as nitrogen) or less, monochloramine is formed. At chlorine ratios of approximately 5:1 to 7:1, dichloramine is formed, and at ratios of 7:1 to 10:1, trichloramine is formed. At ratios above 10:1, all of the ammonia is reacted and free chlorine is again formed. These ratios change with pH and temperature.

Chloramines are generally not effective as a primary disinfectant, because the reactions times required for microbial inactivation are very long. However, chloramines are commonly used to provide residual disinfection in distribution systems. Ammonia-oxidizing bacteria can react with ammonia in the distribution system to cause nitrification, resulting in a loss of residuals and the formation of high levels of heterotrophic bacteria. Careful monitoring of the residual levels and free ammonia present in the water can help control nitrification.

In addition to the chlorine equipment and processes described previously, ammonia systems are needed for chloramination. Ammonia can be fed as a liquid (aqua ammonia) or as a gas (anhydrous ammonia).

INACTIVATION DOSE REQUIREMENTS

Inactivation dose requirements have been established for inactivation of *Giardia*, *Cryptosporidium*, viruses, and other microorganisms. These dose requirements can vary with changing water conditions including the type of disinfectant, the concentration of the disinfectant, pH, and temperature.

Systems Complying With the Groundwater Rule

Systems that must comply with the treatment requirements of the Groundwater Rule are required to implement treatment for 4-log (99.99%) inactivation of viruses. Table 3-3 shows the required CT values for 4-log virus inactivation.

SYSTEMS COMPLYING WITH THE UNFILTERED REQUIREMENTS OF THE SURFACE WATER TREATMENT RULE

Treatment requirements for groundwater systems that must comply with the primary disinfection requirements for groundwater under the influence of surface water may include 3-log (99.9%) inactivation of *Giardia* cysts. This requirement is only for unfiltered systems, and lower log removals may be required if the groundwater is filtered or if credit is received for riverbank filtration. CT and dose requirements for various disinfectants used to inactivate *Giardia* are shown in Table 3-4.

Table 3-3   CT Values Required for 4-log Virus Inactivation Using Various Disinfectants

Chlorine, mg/L min

pH	Temperature, °C					
	0.5	5	10	15	20	25
6.0 to 9.0	6	4	3	2	1	1

Ozone, mg/L min

pH	Temperature, °C					
	1	5	10	15	20	25
6.0 to 9.0	1.8	1.2	1	0.6	0.5	0.3

Chlorine dioxide, mg/L min

pH	Temperature, °C					
	1	5	10	15	20	25
6.0 to 9.0	50.1	33.4	25.1	16.7	12.5	8.4

Chloramine, mg/L min

pH	Temperature, °C					
	1	5	10	15	20	25
6.0 to 9.0	2,883	1,988	1,491	994	746	497

UV dose, mJ/cm²

pH	Temperature, °C					
	1	5	10	15	20	25
6.0 to 9.0	186.0					

Additional treatment may be required for *Cryptosporidium* inactivation for groundwaters under the influence of surface water. Treatment options may include up to 2.0-log (99%) inactivation of *Cryptosporidium* as an additional treatment measure. CT and dose requirements for 2-log *Cryptosporidium* inactivation are shown in Table 3-5.

Table 3-4 CT Values Required for 3-log *Giardia* Inactivation Using Various Disinfectants

Chlorine, mg/L min

pH	Temperature, °C					
	0.5	5	10	15	20	25
6	181	126	95	63	47	32
6.5	217	151	113	76	57	38
7	261	182	137	91	68	46
7.5	316	221	166	111	83	55
8	382	268	201	134	101	67
8.5	460	324	243	162	122	81
9	552	389	292	195	146	97

Ozone, mg/L min

pH	Temperature, °C					
	1	5	10	15	20	25
6.0 to 9.0	2.9	1.9	1.43	0.95	0.72	0.48

Chlorine dioxide, mg/L min

pH	Temperature, °C					
	1	5	10	15	20	25
6.0 to 9.0	63	26	23	19	15	11

Chloramine, mg/L min

pH	Temperature, °C					
	1	5	10	15	20	25
6.0 to 9.0	3,800	2,200	1,850	1,500	1,100	750

UV dose, mJ/cm^2

pH	Temperature, °C					
	1	5	10	15	20	25
6.0 to 9.0	10.8					

Table 3-5 CT Values Required for 2-log *Cryptosporidium* Inactivation With Various Disinfectants

Chlorine, mg/L min

pH	Temperature, °C					
	0.5	5	10	15	20	25
6.0 to 9.0	None established, greater than 4,000					

Ozone, mg/L min

pH	Temperature, °C					
	1	5	10	15	20	25
6.0 to 9.0	46	32	20	12	7.8	4.9

Chlorine dioxide, mg/L min

pH	Temperature, °C					
	1	5	10	15	20	25
6.0 to 9.0	1,220	973	665	357	232	

Table 3-5   CT Values Required for 2-log *Cryptosporidium* Inactivation With Various Disinfectants (continued)

Chloramine, mg/L min

pH	Temperature, °C					
	1	5	10	15	20	25
6.0 to 9.0	3,800	2,200	1,850	1,500	1,100	750

UV, mJ/cm²

pH	Temperature, °C					
	1	5	10	15	20	25
6.0 to 9.0	5.8					

REFERENCES

Chris, W. 2001. The Toxicology of Chlorine. *Environmental Research*, 85:2:105.

Hammond, C.R. 2000. The Elements. *Handbook of Chemistry and Physics*, 81th ed. CRC Press: Boca Raton, Fla.

International Standard Organization. 2007. International Standard 21348ISO, Space Environment (Natural and Artificial)—Process for Determining Solar Irradiances. http://www.spacewx.com/ISO_solar_standard.html. Accessed Aug. 12, 2009.

Kirmeyer, G.J., et al. 1993. *Optimizing Choramine Treatment*. AwwaRF and AWWA: Denver, Colo.

Kirmeyer, G.J., et al. 1995. *Nitrification Occurrence and Control in Chloraminated Water Systems*. AwwaRF and AWWA: Denver, Colo.

NFPA (National Fire Protection Association). 2010. *NFPA 400: Hazardous Materials Code*. National Fire Protection Association: Quincy, Mass.

OSHA (Occupational Safety and Health Administration). 2009. *Code of Federal Regulations*, Title 29, Vol. 6. Occupational Safety and Health Standards. US Government Printing Office: Washington, D.C.

Symons, J.M., et al. 1998. *Factors Affecting DBP Formation During Chloramination*. AwwaRF and AWWA: Denver, Colo.

USEPA (US Environmental Protection Agency). 2006. Ultraviolet Disinfection Guidance Manual for the Long-Term 2 Enhanced Surface Water Treatment Rule (LT2ESWTR). EPA 815-R-06-007.

Valentine, R.L., et al. 1998. *Chloramine Decomposition in Distribution System and Model Waters*. AwwaRF and AWWA: Denver, Colo.

White, G.W. 1999. *The Handbook of Chlorination and Alternative Disinfectants*, 4th ed. John Wiley: New York.

CHAPTER FOUR

Corrosion Control

Corrosion control is required for systems that exceed lead and copper action levels (ALs). The Lead and Copper Rule (LCR) was established in 1989; however, the science and application of corrosion control treatment has advanced far beyond the original guidance provided in the rule. When the LCR was promulgated, corrosion control strategies focused on two areas: minimizing the solubility of lead and copper in water and forming protective scales on the surfaces of distribution and premise plumbing.

The lead AL was set at 0.015 mg/L in the 90th percentile sample of targeted customer plumbing systems. The copper AL was set at 1.3 mg/L in the 90th percentile sample. This new rule meant that utilities had to work with willing customers to sample in-home plumbing systems.

Systems that exceeded the ALs had to develop strategies that would optimize corrosion control treatment without violating other primary health regulations. This was a concern, because many corrosion control strategies focused on pH and alkalinity adjustment to minimize lead and copper solubility. In many cases, adjusting the pH resulted in higher disinfection by-product levels. Consequently, the US Environmental Protection Agency's (USEPA's) strategy allowed water systems to balance water chemistry to minimize lead and copper without causing other violations.

CORROSION CONTROL TREATMENT ALTERNATIVES

An overview of corrosion control strategies is provided in Table 4-1.

Aeration

Aeration can be used to strip carbon dioxide from water systems. This technology's effectiveness is dependent on water quality. Aeration works best for systems with pH below 7.0 but has been used effectively in some waters with pH below 7.5. Aeration will only remove part of the available carbon dioxide from the water, because as the pH rises, some carbon dioxide shifts to bicarbonate species. With an effective air stripping system, pH can be raised to approximately 8.0.

Table 4-1   Corrosion Control Treatment Alternatives

Alternative	Corrosion Control Approach	Forms and Feed Systems	Benefits	Drawbacks
Aeration	Raises pH for waters below pH 8.3	Packed tower aeration, cascade aeration, spray nozzles, bubble diffusion	Nonhazardous; can use ambient air for pH adjustment	Only raises pH by stripping available carbon dioxide from water
Carbon dioxide	Lowers pH; is sometimes added with lime or caustic to increase alkalinity	Gas diffusion in a pipeline or contact chamber	Easy to handle; relatively low cost	Gas transfer efficiency and off-gas must be carefully designed
Lime	Increases pH, alkalinity, and dissolved inorganic carbonate	Pebbled lime must be slaked; hydrated lime in dry form; available as a slurry in some locations	Relatively inexpensive	Chemical feed systems require significant maintenance
Caustic soda	Increases pH and hydroxyl alkalinity	Liquid from 30% to 50% concentration	Can use relatively simple chemical feed system	Hazardous chemical requires careful design and operation
Soda ash	Increases pH and carbonate alkalinity	Powdered form; dry feed systems used	Easy to operate	Chemical feed systems require dust mitigation and significant maintenance
Sodium bicarbonate	Increases carbonate alkalinity	Powdered form; dry feed systems used	Easy to operate	Chemical feed systems require dust mitigation and significant maintenance
Potassium carbonate	Increases pH and carbonate alkalinity	Powdered form; dry feed systems used	Easy to operate	Chemical feed systems require dust mitigation and significant maintenance
Limestone (calcite)	Increases pH and carbonate alkalinity.	Pebble form; used as media in contactor	Easy to operate	Only appropriate for low pH, low to moderate alkalinity waters
Orthophosphate	Reacts with metal surfaces to form	Liquid or powder form; liquid forms include phosphoric acid	Easy to use; effective for lead control in chloraminated waters	Cost varies widely; works better at pH above 7.5

Table 4-1   Corrosion Control Treatment Alternatives

Alternative	Corrosion Control Approach	Forms and Feed Systems	Benefits	Drawbacks
Polyphosphate	Prevents scaling; sequesters iron	Liquid or powdered form	Easy to use	Not as effective as orthophosphate for lead corrosion control
Blended phosphates	Reacts with metal surfaces to form	Liquid or powder form	Easy to use	Cost varies widely
Silicate	Reacts with metal surfaces to form	Liquid form; very viscous	Can use liquid feed systems; increases pH as well as inhibits corrosion	Cost varies widely; viscous; alkaline solution requires careful design of feed systems
Free chlorine	Can provide conditions to form insoluble lead dioxide	Liquid, gas, powder, or pellets	Commonly used for disinfection of distribution systems	Lead dioxide formation requires maintenance of oxidative conditions throughout distribution system

More carbon dioxide is stripped at higher air-to-water ratios (greater than 5:1 or 10:1).

Carbon Dioxide

Carbon dioxide can be used to lower pH and add alkalinity to the water. Carbon dioxide is fed as a gas and diffused into a pipeline or in a contactor tank or basin. Proper gas transfer is needed for an effective system and can be achieved through use of venturi injectors or properly designed diffusers. Carbon dioxide is sometimes added to water in conjunction with lime or caustic soda to add alkalinity to very soft waters.

Lime

Lime can be added as pebbled lime, which must be slaked to hydrolyze the lime, or as powdered or slurried hydrated lime to increase pH, alkalinity, and dissolved inorganic carbonate. Lime slakers require careful temperature and water control to prevent caking of the lime in the slaker. Lime systems require frequent cleaning and maintenance to prevent build-up of lime on the chemical feed lines.

Caustic Soda

Caustic soda is available as a liquid and is added to raise water pH. Caustic soda will also increase the water's hydroxyl alkalinity. It is commercially available in various strengths, but is commonly purchased as a 30% (approximate) solution. The freezing point of caustic soda varies with the strength. At 30% strength, the freezing point is approximately $-1°C$ ($30°F$); at 50% strength, the freezing point is approximately $13°C$ ($57°F$). Careful design and operating precautions are required to safely handle caustic soda.

Soda Ash

Soda ash is added to water to raise pH and to add alkalinity and dissolved inorganic carbonate to water. Soda ash is available as a powder and requires maintenance and cleaning to prevent build-up of precipitated chemical in the feed system and chemical feed lines. Some systems add polyphosphate to the water supply lines to prevent build-up on the feedwater lines and at the point of injection. The solubility of soda ash varies significantly with water temperature. The solubility at $10°C$ ($50°F$) is 14.7%, and the solubility at $25°C$ ($77°F$) is 23%.

Sodium Bicarbonate

Sodium bicarbonate, or baking soda, can be used to add alkalinity to most waters. Sodium bicarbonate will also shift the pH toward a pH of approximately 8.3. Sodium bicarbonate is available in a powdered form; feed systems and operational requirements are similar to those used for soda ash. The solubility of sodium bicarbonate is 7.8% at $18°C$ ($64°F$).

Potassium Carbonate

Potassium carbonate, or potash, is similar to soda ash, with potassium instead of sodium as the cation in the compound. Feed systems and operational requirements are similar to those for soda ash systems. The solubility of potash at $0°C$ ($32°$) is approximately 28%; at $20°C$ ($68°F$) the solubility is approximately 38%.

Limestone (Calcite)

Limestone, or calcite, can be used to raise the pH and to add alkalinity and dissolved inorganic carbonate in low-pH and low-alkalinity waters. Limestone contactors are used to hold media; water dissolves the limestone slowly as it passes through the media bed. Limestone contactor design should provide a velocity slow enough to dissolve the limestone in order to achieve the designed pH (maximum pH with these systems is approximately 8.3). Access to periodically measure and refill the media bed is necessary. If the system is designed in a down-flow mode, the bed must be periodically backwashed.

Orthophosphate

Orthophosphate can be fed as a liquid or powder to provide corrosion control. Orthophosphate reacts with the metal pipe surface to form metallic phosphate compounds, which reduce leaching of metals into the water. Typical doses are 2–8 mg/L. Orthophosphate can be fed in one of several commercial forms or as phosphoric acid, which becomes orthophosphate in water.

Polyphosphate

Polyphosphate can be fed as a liquid or powder. Powders are typically mixed in batches and fed as a solution into the water supply. Polyphosphate is generally associated with sequestering low levels of iron and manganese but has been used successfully to control corrosion of copper in some cases.

Blended Phosphates

Many commercial blends of poly- and orthophosphates are available commercially as liquid and powdered solutions.

Silicate

Silicate solutions have been used for corrosion control with a variety of water qualities. Sodium silicate is available as a liquid in varying strengths and in powdered form. Sodium silicate solutions are very alkaline (pH 11 to 13.3, typically) and very viscous. Sodium silicate solutions have a freezing point that is close that of water; if allowed to freeze, they become glassified and containers or piping must be replaced. Careful design and operation are required for these systems.

Free Chlorine

Based on recent information regarding the formation of insoluble lead dioxide (PbO_2), free chlorine can be used to maintain an oxidation state in the distribution system that can promote the formation of this compound. This chapter discusses the formation of lead dioxide in more detail later.

REGULATORY ISSUES

USEPA's *Lead and Copper Rule Guidance Manual*, published in 1994, evaluated options for helping water systems minimize lead and copper solubility or form protective scale on the surface of pipe material to prevent lead and copper from entering the water.

Lead and copper entering drinking water from household plumbing materials such as pipes, lead solder, and faucets containing brass or bronze can be controlled by changing water quality characteristics. The characteristics that were expected to have the greatest effect on lead and copper corrosion were pH, dissolved inorganic carbonate (DIC), orthophosphate concentration, alkalinity, and buffer intensity. Dissolved oxygen and chlorine residual were also thought to be important considerations for copper. There are many other factors that affect the corrosion of lead and copper, but these factors cannot be easily altered by a water system and were considered to have a lesser effect on corrosion.

One factor that can be adjusted is alkalinity, which is interrelated with pH and DIC and is routinely measured by water systems. Buffer intensity, which is also interrelated with pH and DIC, is another characteristic that was and still is considered very important in maintaining optimal corrosion control and water quality in the distribution system. Buffer intensity is a measure of the water's resistance to changes in pH and is a good indicator of a stable water quality in the distribution system.

After the LCR was adopted and systems implemented treatment, some systems had difficulty in sufficiently reducing lead and copper corrosion. In particular, groundwater systems in the Midwest with neutral pH values, high hardness, and high-alkalinity were having difficulty meeting the copper AL.

As a result, USEPA published a revised LCR guidance manual in 2006 that included the following:

- information on aeration and limestone contactors for corrosion control;
- a listing of the most successful treatment options for copper corrosion control in high-alkalinity/high-DIC waters;

- trade-offs of corrosion control with iron and manganese removal; and
- considerations for corrosion control in light of the new (1994) water quality–based standards for wastewater treatment.

Recently, USEPA researchers and others studying lead exceedances in Washington, D.C., found that oxidation conditions in the distribution system may play an even more meaningful role in the lead levels found in at-the-tap standing samples.

CORROSION INDICES

There are numerous corrosion control indices to help utilities evaluate the potential effect of their water on pipeline materials. A few common indices are described below.

Aggressiveness Index

The aggressiveness index is an indicator of a water's ability to dissolve asbestos-cement pipe, in particular. It is commonly calculated and sometimes used in place of the Langlier saturation index, but is a less precise and more approximate method. It does not incorporate total dissolved solids and temperature impacts in its calculation. The recommended value of greater than 12 indicates water that is not aggressive toward asbestos-cement pipe.

Ryznar Index

The Ryznar index (RI) was developed from empirical observations of corrosion rates in steel mains and heated water in glass coils. The RI is calculated as $2 \times pH_s - pH$, where pH_s is the saturation pH. Water with an RI between 6.5 and 7.0 is considered to be in saturation equilibrium with calcium carbonate. An RI above 7 indicates the water may dissolve calcium carbonate.

Langlier Index

The Langlier saturation index is calculated as $pH - pH_s$. Water with a value greater than 0 is supersaturated with calcium carbonate, values of 0 indicate the water is in equilibrium with calcium carbonate, and values less than 0 indicate water that is undersaturated. This index can be considered a measure of the driving force for deposition or dissolution of calcium carbonate, but it does not predict how much deposition or dissolution will occur.

Calcium Carbonate Precipitation Potential

The calcium carbonate precipitation potential is the theoretical amount of calcium carbonate mass that could precipitate or dissolve on a pipe surface. A recommended value of 4 to 10 mg/L indicates a water that would be likely to provide a protective scale on a pipe surface while minimizing clogging from excessive saturation.

Larson's Ratio

Larson's ratio is the ratio of bicarbonate alkalinity to chloride plus sulfate. Values should be greater than 3 to 5 for iron corrosion control.

KEY WATER QUALITY PARAMETERS FOR CORROSION CONTROL

Following is a discussion of the key water quality parameters that need to be considered when implementing corrosion control strategies.

pH

pH is a measure of a water's acidity or, more specifically, its hydrogen ion concentration. Most drinking waters have a pH in the range from 6 to 10. One common corrosion control treatment strategy is to raise the pH of the source water. This is usually successful for copper corrosion control, in particular (except in high-alkalinity water). pH can be adjusted using chemical or nonchemical processes. At higher pH values, it is thought that there are fewer tendencies for lead and copper to dissolve and enter drinking water, because their solubility was lower at higher pH.

Water pH can vary significantly as water moves through the distribution system, either increasing or decreasing. This change in pH depends on the size of the distribution system, the flow rate, the age and type of plumbing material, as well as water quality. Maintaining sufficient buffer capacity in the water can help prevent distribution system changes in pH.

Alkalinity

Alkalinity is the capacity of water to neutralize acid. It is the sum of carbonate, bicarbonate, and hydroxide anions. Alkalinity is usually reported as milligrams per liter "as calcium carbonate" ($CaCO_3$). Generally, waters with high alkalinities also have high buffering capacities, which resist changes in pH in the distribution system. In low-alkalinity waters, coagulants such as alum of ferric coagulants consume alkalinity as they hydrolyze, resulting in an unstable pH.

Dissolved Inorganic Carbonate

Dissolved inorganic carbonate (DIC) is an estimate of the amount of total carbonates, including carbon dioxide gas, bicarbonate ion, and carbonate ion, in water. It is usually reported as milligrams of carbon per liter (mg C/L). DIC can be calculated from the pH, alkalinity, and ionic strength or, at a minimum, the total dissolved solids concentration of the water. DIC is closely related to the solubility of lead carbonate species, and it can also affect copper solubility, especially in waters with DIC concentrations greater than 30 mg/L.

Buffer Intensity

Buffer intensity is a measure of the resistance of water to changes in pH, either increases or decreases. Bicarbonate and carbonate ions are usually the most important buffering species in drinking waters. At high pH (greater than 9), silicate contributes to buffering. Phosphate can also provide buffering in waters with very low DIC.

Buffering is normally greatest at approximately pH 6.3, decreases toward a minimum at a pH between 8 and 8.5, and then gets increasingly higher as pH increases above 9. Treated waters with a pH of 8 to 8.5 may be subject to variable pH in the distribution network. Variable distribution system pH is even more pronounced in waters that have very low amounts of DIC (less than about 10 mg C/L). Waters with low buffer intensity are prone to pH decreases from uncovered storage, nitrification, corrosion of cast iron pipe, and pH increases from contact with cement pipe surfaces.

Orthophosphate

Orthophosphate (PO_4) can combine with lead and copper in plumbing materials to form several different compounds that usually form an effective corrosion control barrier. The key to ensuring that orthophosphate reduces lead and copper levels were thought to be maintenance of proper pH and orthophosphate residual. For most systems, effective results are seen when a residual of 0.5 to 2 mg/L as P is maintained. At these levels, the solubility of lead carbonate species in water is dramatically less than in water without orthophosphate. This helps to explain the widespread success of orthophosphate for lead reduction under many different water conditions.

When using orthophosphate for lead and copper control, it was thought that the pH had to be maintained within the range of 7.2 to 7.8. If the pH was too low, even high dosages of orthophosphate were thought not to work. New studies indicate that orthophosphate may be effective for lead and copper corrosion control at a pH as low as 6.0. At high pH, poor corrosion-protecting film stability was often observed. Higher concentrations of orthophosphate are often needed to address copper problems compared to those needed for lead.

Chlorine Residual

The addition of chlorine to a groundwater source can aggravate copper corrosion by oxidizing the exposed materials. Chlorine residuals may have a beneficial effect on lead concentrations by converting lead carbonate species to lead dioxide, which is insoluble.

Chloride and sulfate. Chloride and sulfate may cause increased corrosion of metallic pipes by reacting with the metals in solution and causing them to stay soluble or interfering with the formation of film on pipe surfaces. They also contribute to increased conductivity of water. Chloride is about three times as active as sulfate in this regard, but recent studies have shown that the reactions with chloride in water are very complicated and not always detrimental. Pitting of copper is often associated with high concentrations of chloride and sulfate relative to bicarbonate, where high salt concentrations can contribute to the acidification in pits and enhanced conductivity.

Wastewater Discharge

Corrosion control optimization often lowers the concentrations of metals being sent to the wastewater treatment plant (WWTP). WWTPs have limits on metals contained in effluents and sludge. Coordination with the WWTP on metals reduction strategies is recommended prior to implementing corrosion control treatment.

REVISED GUIDANCE MANUAL FOR LEAD AND COPPER

The USEPA current guidance manual includes flowcharts to help systems determine appropriate corrosion control strategies based on their water quality and the best information available at the time. This information is reproduced in Table 4-2.

Table 4-2   USEPA Corrosion Control Treatment Strategies

Exceeded Lead AL	Exceeded Copper AL	Raw Water pH	Iron and Manganese Removal?	DIC	USEPA Recommended Corrosion Control Strategies
Yes	Yes	<7.2	No	<5	Raise pH in 0.5 increments using soda ash, potassium carbonate, caustic and bicarbonate, or limestone contactor, or add orthophosphate
Yes	Yes	<7.2	No	5–15	Raise pH in 0.5 increments using soda ash, potassium carbonate, caustic, or aeration, or add orthophosphate
Yes	Yes	<7.2	No	>15	Raise pH in 0.25 increments using soda ash, potassium carbonate, caustic, or aeration, or add orthophosphate
Yes	Yes	7.2–7.8	No	<5	Raise pH in 0.5 increments and DIC to 5–10 mg C/L; using soda ash, potassium carbonate, caustic and bicarbonate, or limestone contactor
Yes	Yes	7.2–7.8	No	5–25	Raise pH in 0.3 increments using soda ash, potassium carbonate, or caustic, or add orthophosphate
Yes	Yes	7.2–7.8	No	>25	Add orthophosphate
Yes	Yes	7.9–9.5	No	<5	Raise DIC to 5 to 10 C mg/L; using soda ash, potassium carbonate, or sodium bicarbonate
Yes	Yes	7.9–9.5	No	>5	Raise pH in 0.3 increments using caustic
Yes	No	<7.2	No	<5	Raise pH in 0.5 increments and DIC to 5 to 10 mg C/L; using soda ash, potassium carbonate, caustic and bicarbonate, or limestone contactor, or add orthophosphate
Yes	No	<7.2	No	5–12	Raise pH in 0.5 increments using soda ash, potassium carbonate, caustic and bicarbonate, or limestone contactor, or add orthophosphate

AL—action level; DIC—dissolved inorganic carbon

(continued)

Table 4-2   USEPA Corrosion Control Treatment Strategies (continued)

Exceeded Lead AL	Exceeded Copper AL	Raw Water pH	Iron and Manganese Removal?	DIC	USEPA Recommended Corrosion Control Strategies
Yes	No	<7.2	No	>12	Raise pH in 0.25 increments using potassium carbonate, caustic, or aeration, or add orthophosphate
Yes	No	7.2–7.8	No	<5	Raise pH in 0.5 increments and DIC to 5 to 10 mg C/L; using soda ash, potassium carbonate, caustic and bicarbonate, or limestone contactor
Yes	No	7.2–7.8	No	5–25	Raise pH in 0.3 increments using soda ash, potassium carbonate, or caustic, or add orthophosphate
Yes	No	7.2–7.8	No	>25	Add orthophosphate
Yes	No	7.9–9.5	No	<5	Raise DIC to 5 to 10 mg/L C using soda ash, potassium carbonate, or sodium bicarbonate
Yes	No	7.9–9.5	No	>5	Raise pH in 0.3 increments using caustic
No	Yes	<7.2	No	<5	Raise pH in 0.5 increments and DIC to 5 to 10 mg C/L; using soda ash, potassium carbonate, caustic and bicarbonate, or limestone contactor, or add orthophosphate
No	Yes	<7.2	No	5–12	Raise pH in 0.5 increments using soda ash, potassium carbonate, caustic and bicarbonate or limestone contactor, or add orthophosphate
No	Yes	<7.3	No	13–35	Raise pH in 0.25 increments using potassium carbonate, caustic, or aeration, or add orthophosphate
No	Yes	<7.2	No	>35	Raise pH to 7.2–7.8 using aeration and orthophosphate addition
No	Yes	7.2–7.8	No	<5	Raise pH in 0.5 increments and DIC to 5 to 10 mg C/L; using soda ash, potassium carbonate, caustic and bicarbonate, or limestone contactor

AL—action level; DIC—dissolved inorganic carbon

Table 4-2   USEPA Corrosion Control Treatment Strategies

Exceeded Lead AL	Exceeded Copper AL	Raw Water pH	Iron and Manganese Removal?	DIC	USEPA Recommended Corrosion Control Strategies
No	Yes	7.2–7.8	No	5–25	Raise pH in 0.3 increments using soda ash, potassium carbonate, or caustic, or add orthophosphate
No	Yes	7.2–7.8	No	>25	Add orthophosphate
No	Yes	7.9–9.5	No	<5	Raise the pH in 0.3 increments and DIC to 5–10 mg C/L; using soda ash, potassium carbonate, or caustic and sodium bicarbonate
No	Yes	7.9–9.5	No	>5	Add orthophosphate
Either	Either	<7.2	Yes	<5	Raise pH in 0.5 increments and DIC to 5 to 10 mg C/L; using soda ash, potassium carbonate, caustic and bicarbonate, or limestone contactor, or add orthophosphate
Either	Either	<7.2	Yes	5–12	Raise pH in 0.5 increments using aeration, caustic, or sodium silicate
Either	Either	<7.2	Yes	>12	Raise pH in 0.25 increments using, potassium carbonate, caustic, or aeration, or add orthophosphate
Either	Either	>7.2	Yes	<5	Raise DIC to 5 to 10 mg C/L using sodium bicarbonate or sodium silicate
Either	Either	>7.2	Yes	5 to 20	Raise pH in 0.3 increments using soda ash, potassium carbonate, or caustic, or add blended phosphate
Either	Either	>7.2	Yes	>20	Add blended phosphate
Either	Either	<7.2	Have but no removal	<5	Raise pH in 0.5 increments and DIC to 5 to 10 mg C/L; using soda ash or sodium bicarbonate and silicate
Either	Either	<7.2	Have but no removal	5 to 12	Raise pH in 0.5 increments using aeration, caustic, or sodium silicate
Either	Either	<7.2	Have but no removal	12 to 25	Raise pH in 0.25 increments using potassium carbonate, caustic, or aeration, or add orthophosphate
Either	Either	<7.3	Have but no removal	>25	Add orthophosphate

AL—action level; DIC—dissolved inorganic carbon
Source: USEPA (2003)

CORROSION CONTROL TREATMENT APPLICATION CONSIDERATIONS

For some systems, more than one corrosion control treatment option may be chemically viable. Specific treatment criteria, operations complexity, and secondary impacts associated with each treatment option must be considered before making the final selection.

pH, DIC, and Alkalinity Adjustment Systems

Lime, caustic (sodium or potassium hydroxide), soda ash, sodium carbonate, limestone contactors (calcite filters), and aeration (air stripping) are the principal methods for increasing pH, adding DIC, or increasing alkalinity. Carbon dioxide and acid are sometimes used to decrease pH, and carbon dioxide in combination with caustic soda or lime is used to provide additional DIC without raising pH. Sodium carbonate, soda ash, potash, and limestone contactors also increase DIC; aeration decreases DIC. Design considerations for chemical feed systems are provided in Chapter 2. A few additional considerations specific to corrosion control applications are discussed in the following paragraphs.

Lime. Lime is a hazardous chemical used to raise pH, add DIC, and increase the alkalinity of water. It can cause severe burns and damage the eyes. Lime requires careful design and significant operational oversight. Its use is often limited to larger applications.

Caustic soda. Caustic soda, or sodium hydroxide, is a very hazardous chemical used to raise pH. It can cause severe burns and damage the eyes. At a minimum, caustic feed systems should be equipped with an eye-washing system, full shower, eye goggles, protective gloves, boots, aprons, accessible tanks, and chemical containment areas. For very small systems, a safer option such as soda ash should be used if possible.

Soda ash and potash. Soda ash, or sodium carbonate, is a dry chemical that is relatively safe to handle compared to caustic soda. Soda ash increases DIC and pH. Potash, or potassium carbonate, also increases DIC and pH and is relatively safe to use.

Aeration. The pH in aeration systems can be increased by removing carbon dioxide or carbonic acid. Many groundwater systems with low pH strip carbon dioxide from the water using aeration, resulting in a higher pH. Aeration equipment may include spray systems, diffused bubble systems, packed towers, tray systems, and venturi eductors. One disadvantage associated with aeration is that repumping of the water is required. Some states require systems to disinfect the water after aeration.

Limestone contactors. In most instances, a limestone contactor consists of an enclosed filter containing crushed high-purity limestone ($CaCO_3$); however, open gravity flow systems also exist. As the water passes through the limestone, the limestone dissolves, raising the pH, calcium, alkalinity, and DIC of the water. Because the system does not require pumps or continuous addition of limestone or chemical, it is easy to operate and maintain (Figure 4-1). Limestone must be added periodically, and iron and manganese can coat the surface of limestone, reducing its ability to dissolve.

Sodium bicarbonate. Sodium bicarbonate, or baking soda, is used to increase DIC without a large increase in pH. The bicarbonate species will adjust pH to approximately 8.5 but not higher, unless other chemicals or a higher water pH is already present.

IMPACTS ON WATER QUALITY FROM pH ADJUSTMENT

Implementing corrosion treatment or changing corrosion control treatment strategies by adjusting pH can result in changes to water quality. These changes include increased DBP formation, particularly, total trihalomethane compounds, which increase as pH is

Figure 4-1 A calcite contactor can be a cost-effective way to raise pH for groundwater with low pH and low to moderate alkalinity. These calcite contactors are designed for upflow, so backwashing is not necessary.

raised. Haloacetic acid compounds tend not to increase with pH. A higher pH can increase precipitation of iron and manganese compounds and cause increased color. If the water is relatively close to the saturation point for calcium carbonate, increasing the pH may cause this compound to form scale in the water system. Because chlorine is a stronger oxidant at lower pH, corrosion control is usually recommended after primary disinfection.

Phosphate Addition

Orthophosphate or blended ortho- and polyphosphates can be added to a water to control corrosion (Figure 4-2). Orthophosphate inhibitors include zinc orthophosphate, potassium orthophosphate, sodium orthophosphate, and phosphoric acid. Orthophosphate chemicals are available in both liquid and dry forms. Phosphoric acid is not recommended for small systems because it is a strong acid that can be difficult to handle; it is both a skin and inhalation hazard requiring stringent safety procedures.

The orthophosphate portion of blended phosphates is most beneficial for corrosion control; polyphosphate may help to control iron or manganese precipitation. Polyphosphates should be added soon after addition of chlorine in waters with iron and manganese.

Figure 4-2 Phosphate-based corrosion inhibitors can be mixed in solution tanks from dry product or directly fed as a liquid

Water quality impacts. Added phosphates can affect water quality in several ways, including increased scale dissolution and increased algae growth in places where water is used and exposed to sunlight such as fountains.

Silicate Inhibitors

Silicates, which are mixtures of soda ash and silicon dioxide, are used to raise pH and have sequestering capabilities. Consequently, silicates are used by some utilities with low-pH, low-alkalinity water to control corrosion from lead, copper, and iron. Silicates solutions, which are extremely viscous and have a high pH, require special feed pumps, handling equipment, and temperature control.

RECENT INFORMATION ON LEAD AND COPPER CORROSION

The information presented here was developed in large part based on lead and copper solubility for lead carbonate species. In recent years, the following important findings concerning corrosion control have been made:

- In some waters with very low pH, orthophosphate works well to control both copper and lead.
- The oxidation condition of the distribution system is important for lead control.

Solubility of carbonate species is currently used to evaluate lead leaching potential. However, lead solubilities of lead carbonate species in waters without orthophosphate are many times higher than the lead ALs, and optimizing pH and DIC levels can result in a lead solubility well above the lead AL. For example, at pH 9.5 with a DIC of 12 and water with an ionic strength of 0.01, lead carbonate solubility is approximately 0.100 mg/L. The same water with an orthophosphate level of 0.5 mg/L as P has a lead carbonate solubility of less than 0.010 mg/L.

Schock and Giani (2004) and Schock (2007) have shown that the oxidation reduction potential (ORP) of the water distribution system can profoundly impact lead release in water. Under conditions of relatively high ORP in water (but achievable with free chlorine residual), lead dioxide (PbO_2) can be formed. Lead dioxide is virtually insoluble in water and is often found in systems that use free chlorine, especially in groundwater systems. These researchers also showed that if the distribution system pH or ORP drops, lead dioxide can convert to lead carbonate, destabilizing scale and resulting in high lead levels in at-the-tap samples. These findings may lead to new recommendations for distribution system stability and ORP conditions that favor the formation of lead dioxide in systems that continue to have trouble meeting the lead AL.

REFERENCES

AWWA (American Water Works Association). 1999. *Water Treatment Plant Design*, 5th ed. McGraw-Hill: New York.

AwwaRF (American Water Works Research Foundation) and DVGW Foschungstelle. 1996. *Internal Corrosion of Water Distribution Systems.* AwwaRF and AWWA: Denver, Colo.

Clement, J.A., M.R. Schock, and D.A. Lytle. 1994. Controlling Lead and Copper Corrosion and Sequestering of Iron and Manganese. Proc. 1994 ASCE National Conference on Environmental Engineering, Boulder, Colo.

Lytle, D.A. 1994. Experiences with Sodium Silicate in Corrosion Control Studies. 1994 PQ Corp. International Silicate Conference, Valley Forge, Pa., July 19–21.

———. 1994. Silicates for Simultaneous Fe/Mn Stabilization and Corrosion Control. 1994 PQ Corp. International Silicate Conference, Valley Forge, Pa, July 19–21.

———. 2006. Copper Corrosion Update. Ohio AWWA Safe Drinking Water Act Seminar, Columbus, Ohio, November 9.

Lytle, D.A. and M.R. Schock. 1997. An Investigation of the Impact of Alloy Composition and pH on the Corrosion of Brass in Drinking Water. *Adv. Envir. Res.*, 1:2:213.

———. 1998. Using Aeration for Corrosion Control. *Jour. AWWA,* 90:3:74–88. Erratum, *Jour. AWWA*, 90:5:4.

———. 2007. Unsolved Problems With Corrosion and Distribution System Inorganics. AWWA Annual Conference, Toronto, Canada, June 25–28.

Lytle, D.A., M.R. Schock, and J.A. Clement. 1995. Application and Considerations of Aeration for Corrosion Control. AWWA Water Quality and Technology Conference, New Orleans, La., November 12–16.

Lytle, D.A., M.R. Schock, and J. Kempic. 2006. U.S. EPA Lead and Copper Corrosion, Distribution System Research, and Lead Copper Rule Regulatory Update. AWWA Water Quality Technology Conference, Denver, Colo., November 5–8.

Lytle, D.A. and V.L. Snoeyink. 2001. Drinking Water Quality Deterioration in the Distribution System: Colored Water Formation and Its Control. Florida AWWA Region IV Symp., Corrosion Principles and Practices to Optimize Water Quality, Palm Harbor, Fla., September 20.

Sarin, P., J. Bebee, M.A. Beckett, K.K. Jim, D.A. Lytle, J.A. Clement, W.M. Kriven, and V.L. Snoeyink. 2000. Mechanism of Release of Iron from

Corroded Iron/Steel Pipes in Water Distribution Systems. AWWA Annual Conference, Denver, Colo., June 11–15.

Sarin, P., K. Jim, D.A. Lytle, M.A. Beckett, W.M. Kriven, V.L. Snoeyink, and J.A. Clement. 1999. Systematic Study of Corrosion Scales and Their Release of Iron Particles in Drinking Water. AWWA USEPA Particle Measurement and Characterization in Drinking Water Treatment Symp., Nashville, Tenn., March 28–30.

Schock, M.R. 1989. Understanding Corrosion Control Strategies for Lead. *Jour. AWWA*, 81:7:88.

———. 2007. New Insights into Lead Corrosion Control and Treatment Change Impacts (with some considerations towards Cu). Michigan AWWA Section Meeting, Emerging Issues in Water Treatment, Michigan, May 15.

Schock, M.R. and R. Giani. 2004. Oxidant/Disinfection Chemistry and Impacts on Lead Corrosion. AWWA Water Quality Technology Conference, San Antonio, Texas, November 14–18.

Schock, M.R. and D.A. Lytle. 1992. Corrosion Control Principles and Strategies for Reducing Lead and Copper in Drinking Water Systems. Water Quality Association Annual Convention and Exposition, Orlando, Fla., March 11–15.

Sorg, T.J., D.A. Lytle, and M.R. Schock. 1993. Evaluation of Corrosion Inhibitors to Reduce Lead in the Drinking Water in Buildings. Proc. 19th Annual USEPA RREL Hazardous Waste Research Symposium, USEPA Office of Research and Development, Washington, D.C., April.

Stonesifer, K., D.A. Lytle, and M.R. Schock. 2002. Oxidant and Redox Potential Effects on the Corrosion of Distribution System Materials. AWWA Water Quality Technology Conference, Seattle, Wash., November 10–13.

USEPA (US Environmental Protection Agency). 1999. Letterman, R., R.D. Driscoll Jr., M. Haddad, and H.A. Hau. Limestone Bed Contractors for Control of Corrosion at Small Water Utilities. Publication No. EPA/600/2-86/099.

———. 2003. *Final Revised Guidance Manual for Selecting Lead and Copper Control Strategies*. EPA 816-R-03-001. March 2003.

CHAPTER FIVE

Iron and Manganese Removal

Both iron and manganese occur naturally in the environment, and although iron is somewhat more abundant, both occur in groundwater systems. There are no primary drinking water standards for iron and manganese. An estimated 40% of US water utilities exceed the secondary maximum contaminant levels (SMCLs) for iron of 0.3 mg/L and for manganese of 0.05 mg/L. These SMCLs are intended to represent levels above which aesthetic problems can be expected. To prevent long-term aesthetic and operational problems from occurring, others have recommended levels of 0.1 to 0.2 mg/L for iron and 0.01 to 0.05 mg/L for manganese.

Carbon dioxide in groundwater works to dissolve the iron and manganese mineral compounds into the most commonly found forms in water: ferrous bicarbonate ($Fe(HCO_3)_2$) and manganese bicarbonate ($Mn(HCO_3)_2$). Other iron and manganese compounds are also formed in water, most notably, ferrous sulfate, manganous sulfate, and organic iron and manganese compounds.

AESTHETIC AND OPERATIONAL PROBLEMS

Iron and manganese levels found in drinking water are not regulated for purposes of public health. The primary concerns with iron and manganese in groundwater are aesthetic and operational in nature.

Aesthetic Problems

Aesthetic problems can be quite serious. If not removed, iron and manganese will precipitate, given enough time in the distribution system. Even in groundwater systems that do not chlorinate, there is often more than enough dissolved oxygen (DO) to oxidize iron and manganese in the distribution system. (Stoichiometrically, 1 mg/L of DO is enough to oxidize 7 mg/L of iron and 3.5 mg/L of manganese.)

The most commonly reported aesthetic problems with manganese are black or gray stains on plumbing fixtures and laundry. The most commonly reported aesthetic problems with iron are red water, reddish brown stains on plumbing fixtures, and off tastes caused by reactions with the tannic acids in coffees and teas.

Operational Problems

A number of operational problems are associated with iron and manganese in water distribution systems where these minerals often precipitate in water distribution systems. Many utilities report problems in dead-end mains. In addition, iron and manganese precipitates build up inside of distribution pipes where they can be resuspended during periods of high demand or flow reversal.

A troublesome operational problem results from iron and manganese bacteria growth in the distribution system. These bacteria are autotrophic or facultative autotrophs, meaning they derive the energy needed to sustain life and growth by using the small amount of energy given off during oxidation of iron and manganese compounds. These bacteria are often referred to collectively as iron bacteria and include many species such as *Gallionella, Crenothrix, Leptothrix*, and *Sphearotilus*. Many iron and manganese bacteria are sheathed bacteria (at least during part of their life cycle), and the sheath provides a significant amount of protection from chlorine disinfection.

Other operational problems from growth of iron bacteria include significant reduction in pipeline capacity, clogging of meters and valves, discoloration of water, off tastes, and increased chlorine demand. Iron bacteria are controlled by flushing, increasing chlorine residual, and removing iron and manganese at the source. Although increased chlorine residuals may limit the growth of bacteria, complete inactivation can raise chlorine residuals above 50 mg/L for several hours.

TREATMENT ALTERNATIVES

Many water treatment schemes have been developed to remove iron and manganese compounds from water. Systems in use today include
- aeration followed by filtration,
- chlorination followed by filtration,
- ozone followed by filtration,
- chlorine dioxide followed by filtration,
- potassium permanganate followed by filtration,
- biological filtration,
- ion exchange,
- manganese greensand filtration (Figure 5-1),
- oxide-coated sand filtration,
- pyrolusite media filtration,

Figure 5-1 Greensand filter system in California

- membrane filtration,
- stabilization and sequestering,
- lime softening, and
- other combinations or deviations of the technologies listed.

Numerous vendors supply equipment needed for each treatment type. Because much of this equipment is considered proprietary, it can be extremely confusing when selecting the proper system for a particular installation.

Despite the variety of treatment systems available, there are only four removal mechanisms for iron and manganese, and all systems use one or more of these mechanisms. By understanding how the removal mechanisms work, a better understanding of how each iron and manganese system works can be obtained. The four removal mechanisms are

- precipitation,
- adsorption,
- ion exchange, and
- biological uptake.

Each removal mechanism is discussed in the following paragraphs.

Precipitation

Precipitation occurs when a solution is supersaturated with iron or manganese or when iron or manganese has been oxidized. Figure 5-2 shows iron hydroxide (ferric and ferrous) solubility. By increasing the pH to 10 or above, the solubility of ferrous (or dissolved) iron decreases to near 0. In lime softening systems, the pH is raised to reduce the solubility for these metals, and excess concentrations will precipitate.

Iron can also be oxidized from its ferrous state to its ferric state. As shown in Figure 5-3, the solubility of iron at pH above 3 is 0 and, as a result, the iron precipitates. Manganese precipitates in a similar manner, although the solubility of manganous manganese does not approach 0 until the pH is above 11. When manganese precipitates it can form manganese dioxide, which has a solubility of 0 above pH of 6. Because manganese dioxide can form permanganate when further oxidized, careful control of strong oxidants (like ozone) is required.

Oxidation of iron and manganese can result in a precipitate that can be removed by sedimentation or filtration. However, it is unusual to find a sedimentation basin that can effectively remove iron and manganese to finished water levels, although some removal is usually achieved during sedimentation. Filtration is generally more effective at removing the particles after oxidation, but careful filter design is needed. After oxidation, iron and manganese particles are generally 0.2 to 20 μm in size; filtration particles are 500 to 1,500 times larger (0.3 to 1.5 mm depending on the type of media).

Different oxidants can be used to precipitate iron and manganese. It is important to consider the amount needed for each oxidant, the reaction time, and interferences with oxidation. The amount of chemical needed to oxidize iron and manganese is shown in Table 5-1, and the range of reaction times is shown in Table 5-2. Several compounds can interfere with oxidation including hydrogen sulfide, organic compounds, and ammonia. Ammonia exerts a 10:1 demand on chlorine, and organic compounds can affect both the amount of chemical demand and the reaction time by complexing with iron and manganese.

Reactions with oxygen. Although it is frequently reported that the "most common" methods for removing iron and manganese from water are aeration, precipitation, and filtration, this is likely not accurate. First, the reaction rates for manganese oxidation and organically complexed iron oxidation are far too slow to be used economically in water treatment systems. Second, after several years of operation, it is likely that the media used for filtration will develop a coating of iron or manganese oxides that will improve removal and make the reaction catalytic.

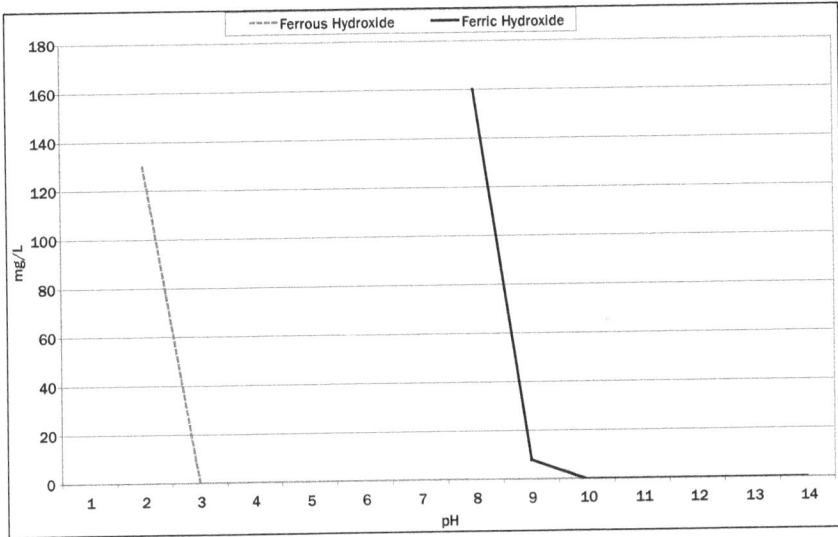

Figure 5-2 Ferrous and ferric hydroxide solubility

(After Werner Stumm and G. Fred Lee, Industrial and Engineering Chemistry, *53:143, 1961)*

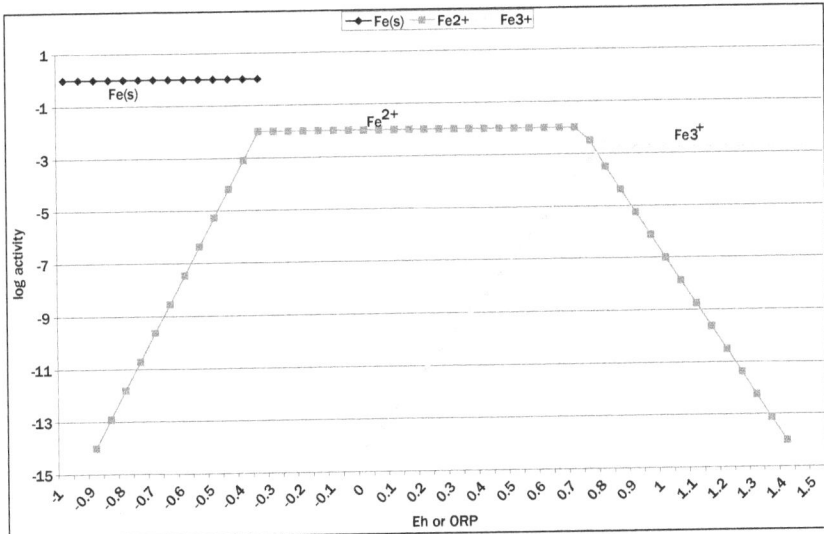

Figure 5-3 Distribution of iron species as a function of oxidation reduction potential

Aeration oxidizes ferrous hydroxide by first stripping off carbon dioxide and then precipitating ferric hydroxide as follows:

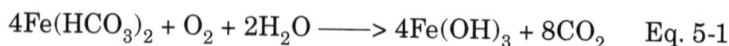

$$4Fe(HCO_3)_2 + O_2 + 2H_2O \longrightarrow 4Fe(OH)_3 + 8CO_2 \qquad \text{Eq. 5-1}$$

Aeration oxidizes manganese as follows:

$$3Mn(HCO_3)_2 + 2O_2 \longrightarrow 3MnO_2 + H_2O + 2CO_2 \qquad \text{Eq. 5-2}$$

The rate of iron oxidation with aeration is slow when pH is below 7. At pH 6.9, approximately 40 min is needed to oxidize 50% of iron in solution. At pH 6.6, less than 10% of iron is oxidized after 50 min. Above pH 7.5, the reaction rate for iron is generally less than 15 min for complete oxidation. In the literature, iron that has been complexed with organic compounds is generally reported as being not readily oxidized by aeration.

The rate of manganese oxidation with aeration is very slow when the pH is below 9. At pH 9.0, approximately 15% of manganese is oxidized after 180 min. At pH 9.3, 50% of manganese is oxidized after 80 min, and at pH 9.5, approximately 50% is oxidized after 40 min. Because of these slow reaction times, aeration alone is generally not used to oxidize manganese.

Table 5-1   Oxidant Requirements for Iron and Manganese

Oxidant	per mg/L of Mn	per mg/L of Fe
Oxygen (from aeration)	0.29	0.14
Ozone	0.67	0.43
Chlorine	1.28	0.63
Potassium permanganate	1.92	0.94
Chlorine dioxide	2.4	1.2

Table 5-2   Oxidation Reaction Times for Iron and Manganese

Oxidant	per mg/L of Mn	per mg/L of Fe
Oxygen (from aeration)	80 min to days, pH-dependent	<1 min to hr, pH-dependent
Ozone	<5 min	<2 min
Chlorine	15 min to 12 hr, pH-dependent	<1 min to 1 hr, pH-dependent
Potassium permanganate	<7 min	<5 min
Chlorine dioxide	<5 min	<5 min

Only 1 mg/L of oxygen is required to oxidize 7 mg/L of dissolved iron, and 1 mg/L of oxygen is required to oxidize 3.5 mg/L of manganese. Because such small amounts of oxygen are needed, even groundwater systems that do not chlorinate and have very low DO levels in the distribution system can expect oxidation and precipitation (given enough time) of iron and manganese.

Reactions with chlorine. Practically speaking, chlorine may truly be the most commonly used oxidant to remove or at least oxidize iron and manganese. Faster reaction times and the widespread use of chlorine as a primary and secondary disinfectant in the United States would indicate that chlorine is used as one of the oxidants in many, if not most, iron and manganese removal systems.

Chlorine generally oxidizes iron fairly rapidly over a wide range of pH. An exception is iron that has been complexed with organic compounds, which requires long reaction times. Iron is generally oxidized by chlorine as follows:

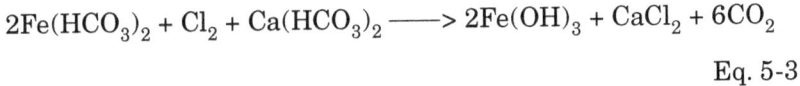

$$2Fe(HCO_3)_2 + Cl_2 + Ca(HCO_3)_2 \longrightarrow 2Fe(OH)_3 + CaCl_2 + 6CO_2$$

Eq. 5-3

Manganese reacts with chlorine more slowly, especially below pH 8.0. The oxidation reaction with manganese is:

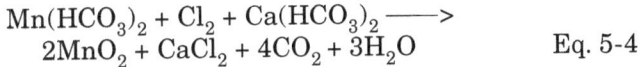

$$Mn(HCO_3)_2 + Cl_2 + Ca(HCO_3)_2 \longrightarrow$$
$$2MnO_2 + CaCl_2 + 4CO_2 + 3H_2O \qquad \text{Eq. 5-4}$$

The reaction rate for chlorine with iron and manganese generally increases with pH. However, the rate of increase is not linear because of the change in chlorine speciation above pH 8 and other factors. Iron oxidation is typically complete with reaction rates well below 1 min, except at very low pH (less than 6.5). Manganese oxidation with chlorine requires nearly 12 hr for completion at pH 6, 2 to 3 hr at pH 8, and 15 min at pH 9. Note that 1 mg/L of chlorine is needed to oxidize 1.61 mg/L of iron, and 1 mg/L of chlorine will oxidize 1.28 mg/L of manganese.

Reactions with ozone. Ozone is a strong oxidant that reacts quickly to oxidize iron and manganese (Figure 5-4). Reactions generally are complete in 5 min in the pH range of 6 to 9. Ozone reacts with iron in the following manner:

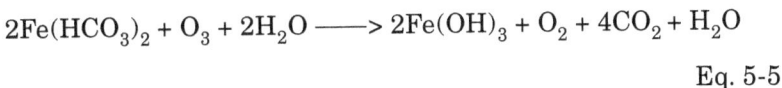

$$2Fe(HCO_3)_2 + O_3 + 2H_2O \longrightarrow 2Fe(OH)_3 + O_2 + 4CO_2 + H_2O$$

Eq. 5-5

Figure 5-4 Ozone contactors and greensand filter at the Camano Water Association in 2005, now replaced with manganese dioxide filters

Ozone oxidation of manganese occurs as follows:

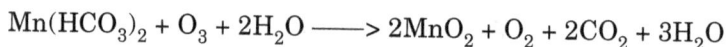

$$Mn(HCO_3)_2 + O_3 + 2H_2O \longrightarrow 2MnO_2 + O_2 + 2CO_2 + 3H_2O$$

Eq. 5-6

A concentration of 1 mg/L of ozone will oxidize 2.3 mg/L of iron and 1.5 mg/L of manganese.

Reactions with potassium permanganate. Potassium permanganate is a strong oxidant that is used extensively in iron and manganese removal facilities (Figure 5-5). Its use includes oxidation of iron and manganese and regeneration of some manganese dioxide–coated media such as greensand. Oxidation reactions with potassium permanganate occur rapidly, typically from 5 to 30 min over a wide range of pH.

Oxidation of iron occurs as follows:

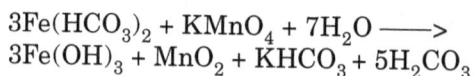

$$3Fe(HCO_3)_2 + KMnO_4 + 7H_2O \longrightarrow$$
$$3Fe(OH)_3 + MnO_2 + KHCO_3 + 5H_2CO_3 \qquad \text{Eq. 5-7}$$

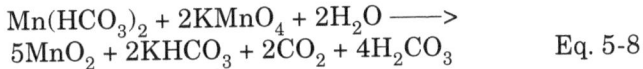

Figure 5-5 Permanganate mixing tank

Oxidation of manganese occurs as follows:

$$Mn(HCO_3)_2 + 2KMnO_4 + 2H_2O \longrightarrow$$
$$5MnO_2 + 2KHCO_3 + 2CO_2 + 4H_2CO_3 \qquad \text{Eq. 5-8}$$

With potassium permanganate, 1 mg/L will oxidize 1.06 mg/L of iron and 0.52 mg/L of manganese.

Adsorption

Iron and manganese are removed by sorption onto oxides on a media surface. Oxides must be in an oxidized state for sorption to occur. Once in this state, the sorption mechanism process occurs very quickly. Chlorine or permanganate is most often used to maintain the oxidation state for these oxides.

Oxides, which can form in a matter of weeks, form naturally on sand or anthracite during iron removal with oxidation. Oxides can also be mined as a mineral, e.g., pyrolusite is marketed as an oxidative media under the trade name AS 741 and others. Oxides can also be formed on media surfaces using permanganate and manganese sulfate. The adsorption capacity is dependent on the oxidation state and the thickness of the coating (in the case of coated media). Silica adsorption on the media surface can cover adsorption sites and reduce effectiveness. This adsorption can often be prevented with permanganate.

The adsorption removal mechanism has been reported to act as an oxidizing contact medium and filtration medium. Adsorption kinetics are much faster than oxidation kinetics. In laboratory tests using manganese concentrations of up to 1.0 mg/L, Knocke (1990) revealed that most uptake occurred in the top 6 in. of the media. This finding was also repeated in full-scale plants at Durham, N.C. Knocke's (1991) later findings included the following:

- The sorption of Mn(II) by MnOx(s)-coated filter media is very rapid. Both sorption kinetics and sorption capacity increase with increasing pH or surface MnOx concentration.
- In the absence of a filter-applied oxidant, Mn(II) removal is by adsorption alone.
- When free chlorine is present, the oxide surface is continually regenerated, promoting efficient Mn(II) removal over extended periods of time.

To maintain efficient uptake kinetics, free chlorine residual in the range of 0.5 to 1.0 is applied as a continuous regenerant. Options for regeneration also include the continuous or periodic application of potassium permanganate.

To take advantage of the rapid reactions that occur with adsorptive removal, systems have been designed with high loading rates (Figure 5-6). Adsorptive removal does not require pH adjustment, contact time for preoxidation, or multiple chemical feeds, unless permanganate is needed to mitigate silicate removal or ammonia oxidation by chlorine. Operating systems with loading rates up to 16 gpm/sq ft have been in operation since 1996 in Washington State.

Ion Exchange

Ion exchange can effectively remove iron and manganese under the right conditions. The removal mechanism for iron and manganese is cation exchange, which is accomplished by exchanging the reduced forms or iron and manganese with sodium on a cation-exchange resin. In their reduced forms, both iron and manganese are divalent cations and have the same charge as calcium and magnesium hardness. The resin must be periodically recharged; this is normally done using a brine solution. Ion exchange does not work well if oxidants are introduced before the ion-exchange resin and may result in resin fouling. The process may also be limited to low levels of iron and manganese and use on soft waters. Some manufacturers recommend

Figure 5-6 These manganese dioxide filters at Clark Public Utilities' Southlake water treatment plant in Vancouver, Wash., will treat up to 10 mgd of groundwater for iron and manganese removal.

that iron concentrations be less than one-tenth of the hardness; others recommend total iron plus manganese of less than 0.5 mg/L.

Biological Uptake

Biological removal of iron and manganese can be accomplished by promoting growth of certain autotrophic bacteria including *Gallionella, Crenothrix, Leptothrix,* and *Sphearotilus* on a media surface. For biological removal of manganese, autotrophic bacteria require specific conditions for effective removal. These bacteria do not require a carbon food source and can extract energy from the oxidation or reduction of inorganic compounds. Oxygen is required to allow growth of iron and manganese bacteria to levels that can efficiently uptake these compounds; however, iron bacteria can live in the absence of oxygen. The specific bacteria that remove iron and manganese are different and grow optimally at slightly different pH and redox conditions. Therefore, they are optimally removed in two separate stages.

Uptake of iron and manganese in biological removal systems can be quite rapid. Both iron and manganese require a period of start-up before bacterial populations reach a density that makes removal effective. After periods of inactivity, a shorter start-up time may also be required. When short-term shutdowns of a few hours occur, iron and manganese removal generally returns to normal after a few minutes. If shutdowns last for several weeks or months, the start-up period may take several hours. Media should be kept wet during shutdowns to minimize the start-up period.

A comparison of iron and manganese removal methods and mechanisms is presented in Table 5-3.

COMMON PROBLEMS AND SOLUTIONS

It is important to know whether the problems within an iron and manganese removal plant occur with iron only, with manganese only, or with both and to know if there has been a sudden decline in performance or if the problem is gradual or seasonal. The common problems that iron and manganese removal plants experience include constant or periodic colored water and poor iron or manganese removal. Many factors contribute to poor iron and manganese removal, and understanding when and how these factors occur is a key to addressing the problem correctly. For example, a plant may see problems only when backwashing occurs, which would point to an issue with hydraulic loading of the filters.

Other common problems include media loss in the filters, changes in water quality, and problems related to improper design. Design problems often occur when a system is not pilot tested. If the process design did not recognize the oxidant amounts, reaction times or particle retention can present issues for the existing system.

Operational problems occur as well. These include under- or over-feeding oxidants, improper backwash rates, and improper backwash frequency and duration. Control systems can unwittingly create operational problems as well. For example, at one Midwestern plant, when the shut down of wells provided water to a greensand plant, the programmable logic controller reset the backwash timer to zero. It took the operator many months to discover that the system was backwashing too infrequently.

In some plants, chemical feeds may not be optimized, either because the doses remained constant and raw water quality conditions changed or because operators reduced chemical feed without initially seeing any decline in performance. In this case, adsorption sites that have developed on the media surface may continue to remove iron and manganese for months. However, these sites will eventually lose their ability to effectively remove iron and manganese.

Table 5-3   Benefits and Drawbacks of Various Iron Removal Technologies

Treatment Technology	Benefits	Drawbacks
Aeration followed by filtration	No chemical use Easy to operate	Entrained air can interfere with filtration if not broken May require breaking head and repumping Not effective for manganese removal or iron complexed with organic material Low filter loading rates for effective removal High capital cost
Chlorination followed by filtration	Chlorine often used for disinfection and present at treatment plant	May require pH adjustment for manganese removal because of slow reactions at low pH Low filter loading rates for effective removal Easy to operate High capital cost
Ozone followed by filtration	Strong oxidant, requires little reaction time	May oxidize manganese to permanganate May oxidize manganese dioxide–containing media to permanganate Difficult to operate High capital and operations and maintenance costs
Chlorine dioxide followed by filtration	Effective for iron complexed with organic material No trihalomethane formation	Generated on site with variety of chemicals Requires careful operation and maintenance Chlorite is a by-product High capital cost
Potassium permanganate followed by filtration	Strong oxidant, requires short reaction times Can reform manganese dioxide coating on media	Causes staining if spilled May be overfed, resulting in pink or purple water
Biological filtration	Easy to operate Low operating cost	Requires start-up period initially and after prolonged shutdowns May require two stages for iron and manganese removal High capital cost

(continued)

Table 5-3   Benefits and Drawbacks of Various Iron Removal Technologies (continued)

Treatment Technology	Benefits	Drawbacks
Ion exchange	Easy to operate	Only effective on reduced forms of iron and manganese No preoxidation should occur before ion-exchange unit Fouling is common Taste may be less palatable than with other methods High capital and operating costs
Manganese greensand filtration	Very effective for manganese Can achieve high loading rates, but often not done	Often used in combination with anthracite media for iron filtration Media may crack Recommended use with permanganate feed
Oxide-coated sand filtration	Naturally occurring or can be man-made on the surface of several types of media Easy to operate	Effectiveness depends on type, thickness, and oxidation state of coating Moderate capital cost
Pyrolusite media filtration	Easy to operate Can achieve high loading rates Low operating costs Very effective for manganese	Moderate capital cost
Membrane filtration	Easy to operate Can achieve high loading rates	May cause fouling Chemical preoxidation must be carefully controlled Moderate to high capital and operating costs
Stabilization, sequestering	May reduce precipitation in parts of the distribution system	Iron and manganese will still precipitate in the distribution system, especially where water stays in the system several days or in hot water systems and appliances Not effective for high levels of iron and manganese
Lime softening	Can effectively precipitate iron and manganese	High capital and operating costs High levels of solids produced Requires significant operational oversight and maintenance

Equipment issues can also contribute to poor performance. Oxidants can impact chemical pump performance by attacking the seats or crystallizing in tubing or injectors. Filters are subject to traditional problems such as media rounding or breakage, mudballs, and air entrainment. Silica and other substances can also coat media and interfere with removal mechanisms, and underdrain breaks can occur undetected.

Plants often use more than one mechanism to remove iron and manganese. For example, if air is introduced into the water prior to filtration, iron may be oxidized quickly and filtered out. The presence of oxygen may also allow the growth of specific autotrophic bacteria that will uptake the iron.

Plants that feed chemicals to oxidize iron and manganese may also find that iron or manganese oxides form on the surface of the media, encouraging adsorption onto the media surface. The first step is to clearly understand the removal mechanisms that are at work or should be at work in the plant. It may not be readily apparent exactly what is happening. Often plants that were designed as oxidation precipitation plants experience development of oxide coatings on the media surface. If air or oxygen is present, biological uptake may be occurring.

A common type of iron and manganese removal plant that uses greensand is supposed to operate through a combination of oxidation/precipitation and adsorption. A process expert should review the plant operation carefully to understand the designed removal process and current conditions.

The second step is to review design parameters. Filter loading rate, media specifications, chemical feed rates, and reaction times should be checked against the original basis of design. Differences should be noted and their impact evaluated.

Media

Media conditions should be evaluated. A core sample can be used to check for breakage, coating, coating damage, mudballs, and rounding. A simple adsorption test can be conducted using a set concentration and volume of manganese standard solution passed through a fixed volume of media. The effluent concentration is measured and the adsorption of the media calculated as milligrams per gram dry media. This is compared to the adsorption capacity of a fresh sample of equal volume using the same media.

Next, raw water conditions should be checked. The raw water should be evaluated for changes from the original design condition. It is not uncommon for iron and manganese levels to increase or decrease over time after pumping. Ammonia, total organic carbon, and hydrogen sulfide may have also increased or decreased, affecting oxidant demand and reaction kinetics.

Equipment

Equipment should be inspected for problems and repaired, if necessary. Pumps should be checked for output against the operating pressure. Valves controlling flow and backwash should be inspected. Filter operation should be monitored at least through one complete backwash cycle. If data are available, a longer period of review is helpful to ensure operations are consistent.

Chemical Feeds

Chemical doses should be checked against the actual demand. Field testing is the most useful way to determine oxidant demand. Simple jar testing equipment with standard solution strengths can be used to add chemicals at multiple doses and the residual measured. The oxidized water can be field-filtered through a 0.25- or 0.45-μm filter to determine if iron and manganese are being precipitated.

Backwashing

Finally, backwashing must be evaluated. The backwashing rate needs to be checked to ensure that the media can be fluidized. The type and size of the media as well as the water temperature affect the required backwash rate, and all must be checked. A build-up of coating on the media can increase its weight, and backwashing rates may need to be increased above the original design values.

Backwash frequency also must be checked. It is helpful to take periodic samples during a filter run to see if performance deteriorates over the run. Backwash duration should also be checked. The backwash should be observed to see if the water clears noticeably by the time the backwash is complete. Samples can be taken and the iron and manganese concentrations in the backwash water can be measured, but dilutions will likely be needed.

A column test is extremely effective at testing potential solutions before large changes in operation or capital facilities are made. Column tests can be run with simple 3- to 6-in.-diameter columns.

The supply water should be identical to the water that is feeding the existing plant. The columns need to be long enough to allow the media bed to expand during backwashing. Clear columns work best, allowing observation of bed expansion during backwashing. Chemical dosing can be optimized, various loading rates can be tested, and media bed designs can be easily changed in a column test. The column test can be done with existing media to test its effectiveness and examine backwash expansion rates or with a new media to test a proposed change.

A comparison of iron and manganese removal technologies is presented in Table 5-3.

REFERENCES

Adams, R.B. 1960. Manganese Removal by Oxidation with Potassium Permanganate. *Jour. AWWA,* 52:2:219.

AWWA (American Water Works Association). 1971. *Water Quality and Treatment*, 3rd ed. McGraw-Hill Book Co.: New York.

Atlas, R.M. 1988. *Microbiology Fundaments and Applications*, 2nd ed. MacMillan Publishing Co.: New York.

Bean, E.L. 1962. Progress Report on Water Quality Criteria. *Jour. AWWA,* 54:11:1313.

Conneley, E.J. 1958. Removal of Iron and Manganese. *Jour. AWWA,* 50:5:697.

Culp/Westner/Culp. 1986. *Handbook of Public Water Systems.* Van Nostrand Reinhold: New York.

Degremont. 1991. *Removal of Iron and Manganese. Water Treatment Handbook*, 6th ed. Paris: Lavoisier Publications, Inc. (Springer Verlag Service Center, Secaucus, N.J.).

Edwards, S.E. and McCall, G.B. 1946. Manganese Removal by Breakpoint Chlorination. *Water and Sewer Work*, 93:303.

Griffin, A.E. and R.J. Baker. 1959. The Breakpoint Process for Free Residual Chlorination. *Jour. New England Water Works Assoc.,* 73:250.

Kawamura, S. 1991. *Integrated Design of Water Treatment Facilities.* John Wiley & Sons, Inc.: New York.

Knocke, W.R., et al. 1990. *Removal of Soluble Manganese from Water by Oxide Coated Filter Media.* AWWA Research Foundation: Denver, Colo.

Morgan, J.J. 1967. Chemical Equilibria and Kinetic Properties of Manganese in Natural Waters. S.D. Foust and J.V. Hunter, eds. Proc. 4th Rudolfs Conference, Principals and Applications of Water Chemistry. John Wiley & Sons: New York.

Mouchet, P. 1992. From Conventional to Biological Removal of Iron and Manganese in France. *Jour. AWWA,* April:158.

Nordell, E. 1951. *Water Treatment for Industrial and Other Uses.* Van Nostrand Reinhold: New York.

Rice, R.G. and A. Nitzer. 1984. *Ozone for Drinking Water Treatment, Handbook of Ozone Technology and Applications,* vol. II. Butterworths Publications: Boston.

Robinson, L.R. and R.I. Dickson. 1968. Iron and Manganese Precipitation in Low Alkalinity Groundwaters. *Water and Sewer Works,* 115:514.

Shair, S. 1975. Iron Bacteria and Red Water. *Ind. Water Eng.,* March/April.

Stumm, W. and G.F. Lee. 1961. Oxygenation of Ferrous Iron. *Ind. Eng. Chem.,* 53:143.

US Executive Department. 1979. National Secondary Drinking Water Regulations, 40 CFR 143. Fed. Reg., p. 42198 (July 19).

White, G.C. 1992. *Handbook of Chlorination and Alternative Disinfectants,* 3rd ed. Van Nostrand Reinhold: New York.

CHAPTER SIX

Arsenic Removal

Arsenic is found in surface water and groundwater as a result of natural processes such as the weathering of minerals and microbial activities. Major anthropogenic sources include mining, particularly smelting, and pesticide manufacture and use. A variety of industries also use arsenic compounds in their production processes.

Generally, naturally occurring arsenic concentrations are below 1 mg/L. Higher concentrations are found in localized zones of contamination. Inorganic forms of arsenic are most common, although organic arsenic compounds associated with microbial activity and pesticide manufacture do occur. Significant concentrations of organic arsenic are generally not found in groundwater used for drinking water supply.

Aqueous, inorganic arsenic can exist in four valence states: As^{+5}, As^{+3}, As, and As^{-3}. Only As^{+5} (also referred to as As(V), or arsenate) and As^{+3} (referred to as As(III), or arsenite) are relevant for drinking water treatment. Speciation is dependent on pH and redox conditions. Oxidizing environments tend to produce arsenate forms; arsenite is produced in acidic-reducing environments. Distribution diagrams for As(III) and As(V) as a function of pH indicate that arsenite is present primarily as the undissociated acid, H_3AsO_3, below pH 9. As(V) is present primarily as $HAsO_4^{2-}$ at pH values above 7; $H_2AsO_4^-$ predominates below pH 7.

TREATMENT ALTERNATIVES

Since the arsenic maximum contaminant level (MCL) was lowered in 2001, there has been considerable research and implementation of full-scale treatment systems for arsenic removal. Several technologies are capable of removing arsenic to low levels. These include conventional and newer technologies such as coagulation, iron and manganese oxidation with precipitation and filtration, lime softening, activated alumina, oxide-coated media, ion exchange, membrane filtration, and electrodialysis reversal. Adsorption onto media containing granular ferric hydroxide, titanium, and aluminum has also been used effectively. Source water quality, operational characteristics, and cost are all factors important to successful application.

Each removal process has advantages and disadvantages, as shown in Table 6-1. Removal mechanisms for arsenic vary with the type of treatment system. The systems listed in Table 6-1 are discussed in this chapter.

Conventional Filtration

The effectiveness of conventional filtration in removing arsenic is dependent on the arsenic species, the type and dose of coagulant used, and the pH of coagulation. Key process considerations for arsenic removal include the following:

- Arsenic has a high affinity for coprecipitation with iron and, to a lesser degree, with aluminum.
- Removal is nearly always better for As(V) than for As(III).
- It is relatively easy to convert As(III) to As(V) with free chlorine.
- Removal declines as pH increases and is limited above pH 8.5.
- Ferric chloride and ferric sulfate generally remove arsenic at lower doses than alum does.
- Silica may interfere with coagulation around pH 7.5.
- Once coprecipitated, arsenic does not tend to leach from solids after drying.

Coagulation can be improved by carefully controlling pH, preoxidizing water prior to coagulation, and adding the proper amount of coagulant. A free chlorine residual of 1.0 mg/L is sufficient to oxidize As(III) to As(V) in 30 to 60 sec for most waters within a pH range of 6 to 9. The disadvantages of conventional filtration are the relatively high capital cost, relatively high operations and maintenance (O&M) costs, and the large amount of solids that must be disposed of properly.

Lime Softening

Lime softening can be used to remove arsenic under the proper conditions. Removal varies with the precipitation of various compounds in the process as follow:

- Magnesium hydroxide precipitation will also remove up to 25% of arsenic.

Table 6-1   Benefits and Drawbacks of Arsenic Removal Technologies

Technology	Benefits	Drawbacks
Conventional filtration	Common technology Effective, especially when arsenic preoxidized and pH kept below 8	Performance declines above pH 8 Arsenic should be preoxidized High coagulant doses sometimes required. Alkalinity addition may be needed for soft waters and high coagulant doses.
Reverse osmosis membrane filtration	Removal of As(III) and As(V) Inorganic, microbial, and organic removal also achieved	Low recovery and flux rates are typical Pretreatment and posttreatment required
Nanofiltration	Removal of As(V) Microbial and organic removal also achieved Removal of calcium and magnesium may be achieved	Sensitivity to water quality Low recovery and flux rates are typical Pretreatment and posttreatment required May not be effective for As(III)
Ultrafiltration	Flux and recovery rates higher than with reverse osmosis or nanofiltration Microbial removal achieved Waste stream can often be sent to wastewater treatment plant	Removal of particulate As only, unless pretreatment with a coagulant is needed for removal Preoxidation and pH adjustment may be needed
Coagulation/ microfiltration	Highest flux and recovery rates of membrane processes Some microbial removal achieved Waste stream can often be sent to wastewater treatment plant	Pretreatment with a coagulant is needed for removal Preoxidation and pH adjustment may be needed
Activated alumina	Less sensitive to water quality than ion exchange Longer run times than ion exchange	pH adjustment often needed Aluminum levels may increase in finished water Hazardous chemicals needed for regeneration Residuals handling is difficult with concentrated high-pH liquid stream
Ion exchange (anion exchange)	Works better at higher pH levels than activated alumina Nitrate removal can also be achieved	Sulfate levels may reduce run times Higher arsenic levels may leach from resin near end of run Requires regeneration and handling of concentrated brine solution

(continued)

Table 6-1   Benefits and Drawbacks of Arsenic Removal Technologies (continued)

Technology	Benefits	Drawbacks
Iron-based sorbents	Arsenic in backwash water is usually very low Relatively easy disposal of solids Some adsorbents have a fairly high sorption capacity	Periodic media replacement required Cost and length of media use before replacement is needed is dependent on water quality Capacity decreases with increasing pH
Titanium-based sorbents	Arsenic in backwash water is usually very low Relatively easy disposal of solids Some adsorbents have a fairly high sorption capacity Works over wide range of pH	Periodic media replacement required Cost and length of media use before replacement is needed is dependent on water quality
Lime softening	Effective removal at pH above 11 Coagulants can be added to aid co-precipitation.	High concentration of solids produced Some systems can require significant operational oversight

- Calcium carbonate precipitation will coprecipitate less than 60% of arsenic.
- Manganese hydroxide precipitation will coprecipitate up to 80% of arsenic.
- Ferric hydroxide precipitate will coprecipitate up to 85% of arsenic (Figure 6-1).

Iron coagulants can be added to the lime softening process to improve arsenic removal. Ferric chloride doses are similar to those used for coagulation. Drawbacks to the softening process include significant O&M needs and a large amount of solids that require proper disposal.

Activated Alumina

Activated alumina removes arsenic by ligand exchange onto an amorphous aluminum oxide. A wide range of effectiveness has been reported for activated alumina, as shown in Table 6-2. The primary measure of effectiveness for activated alumina processes is the number of bed volumes achieved prior to regeneration. Table 6-2 lists three studies in which the number of bed volumes to an arsenic effluent end point ranged from 700 to more than 57,000. The primary

Figure 6-1 The 2003 Juan Plant in Southern California was one of the first US plants to use ferric hydroxide for arsenic adsorption.

reason for the differences in effectiveness is varying water quality. A number of competing ions are removed by this process, as follows:

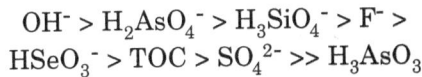

$$OH^- > H_2AsO_4^- > H_3SiO_4^- > F^- >$$
$$HSeO_3^- > TOC > SO_4^{2-} \gg H_3AsO_3$$

The effectiveness of activated alumina treatment can be improved by reducing the pH of the water and thereby reducing the concentration of hydroxyl ions. Effectiveness can also be improved by preoxidizing arsenic before the activated alumina media. Silica, selenium, organic compounds, and sulfate all may contribute to short run lengths between regenerations.

If the water has significant concentrations of these ions, performance of the activated alumina suffers. The alumina media bed is regenerated with a strong-base and weak-acid combination. Drawbacks include disposal of regeneration wastes, high O&M costs, and variable effectiveness. Some sulfur- and iron-modified activated alumina media have been introduced to improve effectiveness.

Ion Exchange

As with the previous three processes, ion exchange is more effective for As(V) removal than for As(III) removal. Ion-exchange run

Table 6-2   Activated Alumina Removal of Arsenic in Three Studies

Influent Arsenic, µg/L	Bed Volumes to Regeneration	Effluent Arsenic Concentration at Regeneration, µg/L	Study
90	700	50	Clifford and Lin, 1995
80–116	8,550	5	Hathaway and Rubel, 1987
22	57,500	10	Simms and Azizian, 1997

Table 6-3   Ion-Exchange Removal of Arsenic in Four Studies

Influent Arsenic, µg/L	Bed Volumes to Regeneration	Effluent Arsenic Concentration at Regeneration, µg/L	Study
88	200	30	Clifford and Lin, 1991
90	4,200	50	Clifford and Lin, 1995
80–116	493	50	Hathaway and Rubel, 1987
21–29	400	<2	Clifford et al., 1998

times before regeneration are generally not long, as with as activated alumina (Table 6-3). Ion exchange does work well under a wide pH range. The effectiveness of ion exchange varies from water to water because of competing ion adsorption. Competing ions include sulfate, nitrate, bicarbonate, and chloride.

Although ion-exchange systems generally work better for As(V) removal than for As(III) removal, prechlorination of water prior to ion exchange is not recommended. Chlorination of resins can produce nitrosamine compounds including NDMA (N-nitrosodimethylamine). Nitrosamines are not federally regulated; however, recent health effects studies have shown nitrosamines to be carcinogenic, teratogenic, and mutagenic in animal tests, and they are currently regulated in the state of California.

Membrane Processes and Electrodialysis Reversal Systems

Membrane processes vary in their effectiveness in removing arsenic. As shown in Figure 6-2 and Table 6-4, reverse osmosis will

Table 6-4   Membrane Performance for Arsenic Removal

Membrane Type	# Tested	As(V) Removal, %	As(III) Removal, %
Groundwater			
RO	1	86 to >94	N/A
NF	1	62 to 89	N/A
UF	1	34 to 72	N/A
Surface Water			
RO	4	>96	51 to 80
NF	3	>96	20 to 44
UF	1	47	7

Source: Brandhuber and Amy, Water Research Foundation.
N/A—not available from this study; NF—nanofiltration; RO—reverse osmosis;
UF—ultrafiltration

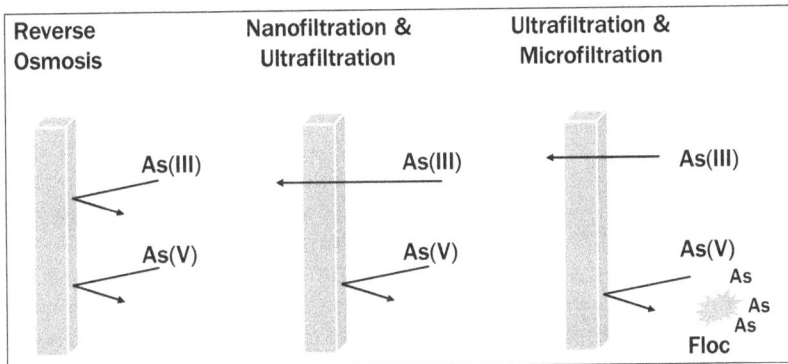

Figure 6-2 Schematic of arsenic removal with various membranes

generally remove arsenate, As(V), and arsenite, As(III). Ultrafiltration andnanofiltration have been used with some success to remove arsenate, but removal of arsenite is poor. Microfiltration will only remove coagulated or particulate arsenic. With microfiltration and coagulation, effective removal can be achieved in manner similar to that of coagulation–filtration. The keys to effective arsenic removal with microfiltration membrane systems are as follows:

- Removal is better for As(V) than for As(III).
- It is easy to oxidize As(III) to As(V); however, the membrane material needs to be compatible with the oxidant used.

- Removal effectiveness improves at lower pH and is limited above pH 8.5.
- Coagulant dose is important, and ferric chloride or ferric sulfate typically works better than aluminum salt coagulants.
- Membrane pore sizes of 0.2 µm and 0.45 µm work better than the 1-µm pore size.

The drawbacks of membrane systems include relatively high capital cost and disposal of backwash and cleaning wastes that contain elevated arsenic levels. These wastes are sent to the sanitary sewer, if possible.

Adsorption Systems

The lower MCL set for arsenic spurned a new market for adsorptive treatment media. Some early media had limited success, but many more are working well. Initial predictions of run-times for many of these media were optimistic, and costs have not decreased for most media in the last 10 years. A number of new sorbents are being used to remove arsenic (Table 6-5). Many of these products have been developed within the last few years and there is limited operational data.

Nearly all, if not all, of these adsorbents perform better for As(V) compounds than for As(III) compounds (Figure 6-2). However, some adsorbent media are formed with organic resin bases and could form nitrosamine compounds if prechlorination is used. Natural organic matter in raw water can decrease the effectiveness of adsorbent media. The raw water pH is also very important for the performance of most adsorbent media, with adsorption of arsenic decreasing as pH increases. The exception to this may be titanium-based adsorbents, which seem to work well over the pH range of 5 to 10. Other important considerations for adsorptive media used for arsenic removal include the following:

- Expected life of the media is highly variable based on water quality. Parameters that affect performance vary with each type of adsorptive media. Extensive water quality information should be provided to equipment suppliers or rapid, small-scale column tests should be performed before selecting an adsorptive media.

Table 6-5   Comparison of Arsenic Removal Sorbents

Media	Initial Arsenic, µg/L	Water Source	BV to 10 µg/L AU//is BV=bed volume	mg As Absorbed per g Media	g Iron per g Media	Source
Iron–citric acid preloaded GAC	50–60	Rutland, Mass. pH 6	150,000	4.96	0.0054	AwwaRF, 2007
Ferrichite (FeCl₃ + chitosand)	3,580	Superfund Tacoma, Wash.	700	1.1	0.61	Chen et al., 2000
Chemical coating onto absorption media G2	200	Spiked distilled water	5,000	2	Iron content not available	Winchester et al., 2000
Granular ferric hydroxide; Wasserchemie	16	Wildeck, Germany	85,000–7 µg/L	0.82	0.58	Driehaus, 2000
Granular ferric hydroxide	21	Stadtoldentrof, Germany	75,000–7 µg/L	1.08	0.58	Jekel and Seith, 2000
Granular ferric oxide media; US Filter/Siemens	18	Stockton, Calif.	25,000	0.2	0.58	McAuley, 2004
Granular ferric oxide media; Severn Trent	18	Stockton, Calif.	25,000	0.2	0.63	McAuley, 2004
Granular ferric oxide media; Wasserchemie	8	Barkersfield, Calif.	80,000–4 µg/L	0.26	0.58	McAuley, 2004
Granular ferric oxide media; Severn Trent	8	Barkersfield, Calif.	80,000–4 µg/L	0.26	0.63	McAuley, 2004

(continued)

Table 6-5   Comparison of Arsenic Removal Sorbents (continued)

Media	Initial Arsenic, µg/L	Water Source	BV to 10 µg/L AU//is BV=bed volume	mg As Absorbed per g Media	g Iron per g Media	Source
Granular ferric oxide media; Wasserchemie and US Filter/Siemens	15	Deionized water spiked with As	60,000–7 µg/L	0.58	0.58	Bradruzzaman et al., 2001
Zirconium-loaded activated carbon	500	Carbonate buffer spiked with As	5,900	2.8	0.028 g Zr/g	Daus et al., 2004
Absorptionsmittel 3 (UFZ-Umwelt-forschungszentrum Leipzig-Halle GmbH)	500	Carbonate buffer spiked with As	1,000	2	0.075	Daus et al., 2004
Iron hydroxide granules	500	Carbonate buffer spiked with As	13,100	2.3	0.323	Daus et al., 2004
Iron-impregnated polymer resin	50	Deionized water with anions, pH 7.5	4,000	0.32	0.09–0.12	DeMarco et al., 2003
Iron oxide–impregnated activated alumina	500	Deionized water with As, pH 12	500–50 µg/L	0.29	0.066	Kuriakose et al., 2004

Source: Chen et al., 2007.

Table 6-6   NSF Listed and USEPA ETV Arsenic Removal Media

Media	NSF	USEPA ETV
Alcan Chemicals' Actiguard AAFS50 (Montreal, QB, Canada)	✓	✓
ADI International, Inc. G2, G2-R (Fredrickton, NB, Canada)	✓	✓
Siemens Water Technologies Corp. GEH, GFH (Berlin, Germany)	✓	
Water Remediation Technology (WRT), LLC. Z-33, Z-88, Z-88AM (Wheat Ridge, Colo.)	✓	
Kinetico Incorporated, Ultrasorb T (Newbury, Ohio)	✓	

USEPA ETV—US Environmental Protection Agency Environmental Technology Verification Program

Table 6-7   Types of Residuals From Arsenic Treatment Technologies

Treatment	Type of Residual	Characterization of Residual
Ion exchange	Brine	High TDS waste stream; not settleable; may leach arsenic
Activated alumina	Concentrated liquid	High pH and low pH waste streams with elevated arsenic concentrations; not settleable
Reverse osmosis	Brine	High TDS waste stream; not settleable
Nanofiltration	Concentrated liquid	High mineral waste stream; possibly settleable
Ultrafiltration	Concentrated liquid	Settleable waste stream from backwashing; concentrated cleaning waste not settleable
Microfiltration	Concentrated liquid	Settleable waste stream from backwashing; concentrated cleaning waste not settleable
Iron and manganese treatment systems	Concentrated liquid	Settleable waste stream
Coagulation–filtration	Concentrated liquid	Settleable waste stream
Lime softening	Sludge	Settleable waste stream
Adsorptive media	Solids	Solid media

TDS—total dissolved solids

- Municipalities may want to design vessels for media contacting and purchase them separately from the media supplier to take advantage of competitive bidding when media need to be replaced in the future.

To date, only a few adsorptive media have completed NSF certification and/or the US Environmental Protection Agency's Environmental Technology Verification (ETV) Program (Table 6-6).

Iron and Manganese Removal Systems for Arsenic

Several studies have shown the arsenic removal capabilities of manganese dioxide–coated media (Figure 6-3). Removal of arsenic occurs by coprecipitation and removal with iron in these systems. The oxidants used are often effective at oxidizing As(III) to As(V), thus improving performance. Important considerations for these types of systems include the following:

- Preoxidation to form As(V) is recommended and is often compatible with iron and manganese removal.
- If sufficient iron needs to be present in the raw water source to remove arsenic, ferric coagulant addition may be needed.
- Arsenic removal works better at lower pH, although oxidation rates for iron and manganese generally slow at lower pH.
- Careful media bed design is needed to retain precipitates.
- Backwashing frequency may need to be shortened in existing iron removal systems if ferric coagulants are added.
- Once coprecipitated with iron, arsenic does not tend to leach from backwash solids.

HANDLING AND DISPOSAL

Residuals handling and disposal options often drive decisions on which type of arsenic system to use. The types of residuals that must be dealt with vary depending on the process used. A summary of residual types and characteristics for arsenic removal systems is provided in Table 6-7. Most residuals from systems using iron or aluminum for coprecipitation with arsenic (ultrafiltration, microfiltration, coagulation–filtration, iron removal, and lime softening systems) are relatively easy to dispose of. The residuals from backwashing these systems can be settled in a tank or lagoon before dewatering and drying. Once dried, they can be shipped to a landfill.

Figure 6-3 This groundwater treatment plant uses preoxidation and ferric chloride addition to remove arsenic in manganese dioxide filters.

Liquid streams that do not readily settle are more difficult to handle. If available, sanitary sewer disposal may be an option, although wastewater treatment plants have limits on discharge and land application for arsenic. Mechanical dewatering, precipitation, and evaporation ponds are potential residuals handling alternatives for these systems.

Toxicity contaminant leachate potential (TCLP) testing is used to determine the suitability of dewatered solids for shipment to a landfill. Federal guidelines for the TCLP limit arsenic in the leachate to a maximum of 5 mg/L. Most residuals coprecipitated with iron or alumina coagulants and most adsorptive media do not exceed 5 mg/L for arsenic in the TCLP test and can be sent to a nonhazardous waste location.

REFERENCES

Amy, G.L., M. Edwards, M. Benjamin, K. Carlson, J. Chwirka, L. Brandhuber, L. McNeill, and F. Vagliasindi. 1999. *Arsenic Treatability Options and Evaluation of Residuals Management Issues.* Awwa Research Foundation: Denver, Colo.

AWWA (American Water Works Association). 1999. *Water Quality and Treatment*, 5th ed. McGraw-Hill: New York.

Benjamin, M. and J. Morgan. 1998 Sorption of Arsenic by Various Adsorbents. AWWA Inorganics Contaminants Workshop, San Antonio, Texas (February).

Brandhuber, Philip and C. Amy. 2000. Identification of Key Engineering Parameters Influencing the Treatment of Arsenic in Drinking Water Via Membrane Technology. In *Proc. of 2000 Inorganic Contaminants Workshop*. AWWA: Denver, Colo.

Chen, W., R. Parette, J. Zou, F. S. Cannon, and B. A. Dempsey. 2007. Arsenic Removal by Iron-Modified Activated Carbon, Draft Project Report #3158. Water Research Foundation: Denver, Colo.

Cheng, R.C., S. Liang, H. Wang, and J. Beuhler. Enhanced Coagulation for Arsenic Removal. *Jour. AWWA*, 9:79.

Chowdhury, Z., S. Kommineni, and Y. Chang. 2002. Comparison of Innovative Technologies for Arsenic Removal. AWWA Inorganics Contaminants Workshop, San Diego, Calif.

Chowdhury, Z., S. Kommineni, R. Narasinhan, J. Brerton, G. Amy, and S. Sinhan. 2002. *Implementation of Arsenic Treatment Systems, Process Selection*. AWWA Research Foundation: Denver, Colo.

Chwirka, J., B. Thomson, and J. Stomp, III. 2000. Removing Arsenic from Groundwater. *Jour. AWWA*, 92:3:79.

Clifford, D.A., and C.C. Lin. 1991. Arsenic (III) and Arsenic (V) Removal from Drinking Water in San Ysidro, New Mexico. Cincinnati, Ohio: US Environmental Protection Agency.

————. 1995. Ion Exchange, Activated Alumina, and Membrane Processes for Arsenic Removal from Groundwater. 45th Envir. Engrg. Conf., University of Kansas.

Clifford, D.A., G. Ghurye, and A. Tripp. 1998. Arsenic Ion Exchange Process with Reuse of Spent Brine. Proc. Annual AWWA Conference. Denver, Colo.

Driehaus, W., M. Jekel, and U. Hildebrandt. Granular Ferric Hydroxide—A New Adsorbent for the Removal of Arsenic from Natural Water. *Jour. Water SRT–Aqua*, 47:1:30.

Edwards, M.A. 1994. Chemistry of Arsenic Removal During Coagulation and Fe-Mn Oxidation. *Jour. AWWA*, 85:9:64.

Hathaway, S. and F. Rubel. 1987. Removing Arsenic from Drinking Water. *Jour. AWWA*, 79:8:61.

Hering, J.G., P. Chen, J. Wilkie, and M. Elimelich. 1997. Arsenic Removal by Ferric Chloride. *Jour. AWWA*, 88:4:155.

Simms, J. and F. Azizian. 1997. Pilot Plant Trials on the Removal of Arsenic from Potable Water Using Activated Alumina. AWWA Water Quality Technology Conference.

USEPA (US Environmental Protection Agency). 2000. Arsenic Removal from Drinking Water by Ion Exchange and Activated Alumina Plants. EPA # 68-C7-0008.

————. 2000. Arsenic Removal from Drinking Water by Iron Removal Plants. EPA # 68-C7-0008.

———.2000. Treatment of Arsenic Residuals from Drinking Water Removal Processes. EPA # 68-C7-0008.

———.2000. Arsenic Removal from Drinking Water by Coagulation/Filtration and Lime Softening Plants. EPA # 68-C7-008.

Vagliasindi, F.G.A. and M. Benjamin. 2001. Redox Reactions of arsenic in experimental solutions and effects on its adsorption. *Jour. Water Supply: Research and Technology–Aqua,* 50:4.

CHAPTER SEVEN

Hydrogen Sulfide Removal

Hydrogen sulfide, which occurs in many groundwaters, is formed by sulfur and sulfate-reducing bacteria that can occur naturally in water. These anaerobic bacteria use sulfates and sulfur compounds found in decaying plant material, rocks, and soil to convert organic compounds into energy. Under these anaerobic conditions, hydrogen sulfide forms as a by-product.

In natural water, sulfur exists in five common stable forms: bisulfate (HSO_4), sulfate (SO_4), thiosulfate ($H_2S_2O_3$), hydrogen sulfide (H_2S), and bisulfide (HS). Other species exist; however, they are not thermodynamically stable. In waters with a normal pH of 8 or below, H_2S and HS are the dominant forms of sulfur, although ionized forms of hydrogen sulfide readily exist in this pH range. The H_2S form becomes more predominant as pH decreases. At pH levels of 8 and above, the reduced sulfur exists in the water as HS and SO_4 ions and the amount of free H_2S is very small.

The US Environmental Protection Agency (USEPA) does not regulate hydrogen sulfide. It is presumed that concentrations high enough to be a drinking water health hazard also make the water unpalatable. The odor of water with as little as 0.5 mg/L of hydrogen sulfide concentration is detectable by most people. Concentrations less than 1 mg/L give the water a "musty" or "swampy" odor. A concentration of 1 to 2 mg/L gives water a "rotten egg" odor and can increase corrosivity in plumbing materials.

HYDROGEN SULFIDE TREATMENT ALTERNATIVES

The recommended treatment for removing hydrogen sulfide from a water supply depends largely on the concentration. The most common method has been aeration and/or oxidation. Activated carbon and iron and manganese adsorption filters have also been used.

The primary problem with each of these removal techniques is that either they do not completely remove the hydrogen sulfide or they form intermediate compounds called polysulfides. Polysulfides can result in off-tastes that have been described as "chemical," "rubber tire," and "musty." These tastes can be made worse when the water is heated.

Oxidation

Chlorine, permanganate, ozone, and chlorine dioxide all oxidize hydrogen sulfide. The major concerns with oxidation are twofold. First, the oxidation process may take a long time to occur and often intermediate compounds have off-tastes. Second, if the distribution system is not maintained in an oxidative state (if chlorine residual is lost or if water stays in a customer's hot water tank for an extended period of time), the sulfide compounds can revert back to their hydrogen sulfide form.

Chlorination. Continuous chlorination is a very common, partially effective method for oxidizing hydrogen sulfide. The recommended dosage is 2.0 mg/L chlorine for every 1.0 mg/L hydrogen sulfide. Chlorine oxidizes hydrogen sulfide to polysulfides and, after a period of time, to sulfate. Depending on the pH and temperature of the water, complete oxidation to sulfate may take several days. Intermediate sulfide compounds can impart metallic tastes to the water, and these compounds may revert back to hydrogen sulfide in areas of no chlorine residual, long contact times, or elevated temperature.

Oxidation/Reduction

One effective, but little known practice for removing hydrogen sulfide is a two-step chemical reaction that oxidizes hydrogen sulfide to polysulfides and then reduces the polysulfides and elemental sulfur to thiosulfate and sulfate. The main advantage of this system is that the reactions take place very quickly and can often be completed without the use of contact tanks. Sulfur dioxide, which is the most commonly used reducing agent for this purpose, is applied as a gas using equipment similar to gas chlorination equipment. Sodium bisulfite and ascorbic acid can also be used to complete the reaction. These chemicals are added as a solution. Complete mixing must be provided at each step, and reducing chemicals must be carefully controlled to prevent a loss of disinfectant residual in the distribution system.

Adsorptive Media

Adsorptive media, including greensand, pyrolusite, and granular activated carbon, have been used to reduce hydrogen sulfide.

Manganese greensand. Manganese greensand has been used with some success for more than 50 years to remove sulfur from drinking water. It is usually recommended for water that contains less than 5.0 mg/L hydrogen sulfide. A manganese greensand filter has

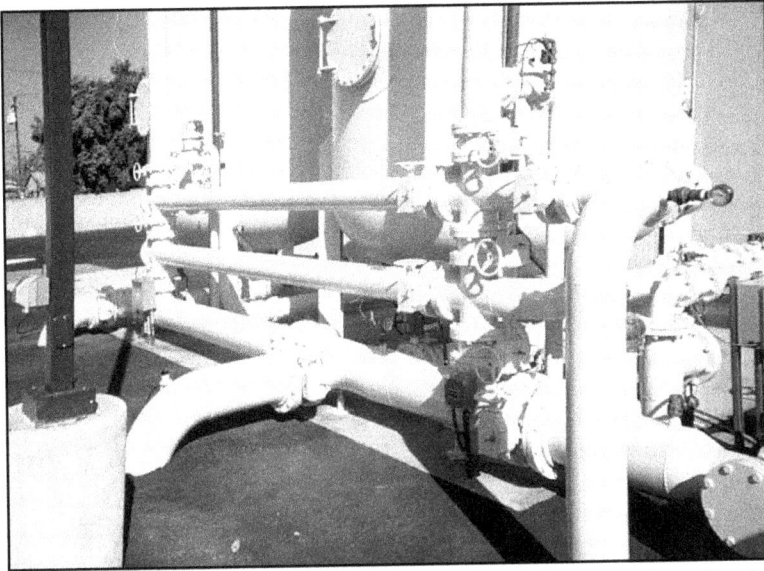

Figure 7-1   These GAC contactors use coconut shell granular activated carbon and air to remove hydrogen sulfide from groundwater.

a manganese dioxide coating that catalyzes hydrogen sulfide gas to solid sulfur particles, which are then filtered. Greensand is regenerated, or recoated, with potassium permanganate or chlorine. A prechlorination step is recommended to oxidize hydrogen sulfide and help regenerate the manganese greensand filter.

Pyrolusite. Pyrolusite, the mineral form of manganese dioxide, has been used for hydrogen sulfide, iron, and manganese removal. Adsorption must be maintained in the media bed by applying a free chlorine residual of 0.5 mg/L or more across the bed.

Granular activated carbon. Although several methods are available for treating hydrogen sulfide in drinking water, advancements in the use of granular activated carbon provide an effective alternative to chemical treatment(Figure 7-1). For hydrogen sulfide removal, the activated carbon is modified with a carbon surface and called catalytic carbon.

Granular activated carbon that has not been modified will remove small amounts of hydrogen sulfide, generally to concentrations below 0.3 mg/L. Catalytic carbon retains all of the adsorptive properties of conventional activated carbon but combines them with the ability to promote or catalyze chemical reactions. During the treatment process, catalytic carbon first adsorbs sulfides onto the carbon surface. Then, in the presence of dissolved oxygen, it oxidizes the sulfides and converts them to nonobjectionable compounds.

Several design considerations affect the performance of catalytic carbon, including empty bed contact time (typically 5 min or longer), backwash capability (backwashing with treated water is recommended to remove any solid or filtered material such as elemental sulfur), and the concentrations of hydrogen sulfide and dissolved oxygen in water. A minimum dissolved oxygen level of 4.0 mg/L is necessary for complete oxidation of hydrogen sulfide to elemental sulfur.

Aeration

Another common treatment for sulfur in water is aeration. Hydrogen sulfide is physically removed by agitating the water via bubbling or cascading and then separating, or "stripping," the hydrogen sulfide in a container. The hydrogen sulfide is removed as a volatile gas by venting it into a waste pipe or to the outdoors. Aeration is most effective when hydrogen sulfide concentrations are below 2.0 mg/L.

The pH of the water plays a significant role in the effectiveness of hydrogen sulfide removal. At pH 6, roughly 80% can be removed by aeration, at pH 7 only about 30% can be removed, and at pH 8, less than 10% can be removed effectively with aeration. Reduction in pH with aeration is often required.

Preoxidation is not recommended with aeration of hydrogen sulfide. Preoxidation may produce sulfide, bisulfide, or solid sulfur particles, all of which are not air-strippable and need to be filtered from the treated water. If not removed, they may revert back to hydrogen sulfide if reducing conditions are present. Hydrogen sulfide can promote growth of bacteria in aeration systems and equipment, requiring periodic cleaning. Since aeration is usually practiced at atmospheric pressure, treated water must be repumped after aeration for service at distribution pressures.

Ion Exchange

Ion exchange works by exchanging a chemical or contaminant on a resin column for another, less objectionable chemical or contaminant. In general, two types of ion exchange exist: cation exchange and anion exchange. Cation-exchange units remove positively charged constituents, such as the hardness minerals calcium and magnesium, and replace them with sodium or potassium. Anion-exchange

Table 7-1   Alternatives for Hydrogen Sulfide Removal

Treatment	Benefits	Drawbacks
Catalytic carbon–granular activated carbon	Effectively controls hydrogen sulfide tastes with proper carbon selection	Carbon must be replaced periodically Dissolved oxygen level of 4 mg/L or greater is needed Moderate capital cost and moderate to high operating cost
Greensand	Reduces tastes and odors Low operating cost	Moderate capital cost Media must be regenerated Media subject to cracking at high head loss
Pyrolusite	Reduces tastes and odors Low operating cost	Moderate capital costs Requires chlorine residual on media bed
Ion exchange	Effectively controls hydrogen sulfide tastes with proper selection and maintenance	Requires salt regeneration High capital and operations and maintenance costs
Chlorination	Reduces hydrogen sulfide smell	Generates polysulfides, which also have tastes and odors Can revert to form hydrogen sulfide if reducing conditions exists (dead-end mains, customer hot water tanks)
Aeration	Reduces tastes and odors	Must repump water after aeration May require acid feed to lower pH and improve effectiveness High capital cost Moderate operating cost
Oxidation/reduction	Effectively controls hydrogen sulfide tastes with proper design and maintenance	Requires second chemical feed Requires effective blending and reaction period Reducing chemical dose must be carefully controlled

units remove negatively charged constituents, such as nitrate and sulfate, and replace them with chloride. Some mixed media ion-exchange units remove both cations and anions.

Hydrogen sulfide can be removed using anion-exchange resins, because a significant amount of hydrogen sulfide present in water is ionized. The effectiveness of the system depends on the resin selected and the concentration of competing anions (sulfate, total organic carbon, alkalinity), as well as the pH of the water. Commercially available strong-base resins in the chloride form are used to remove hydrogen sulfide. Regeneration, backwashing, and rinsing are no different than in other applications (nitrate or arsenic). Table 7-1 provides a comparison of each method used to remove hydrogen sulfide.

REFERENCES

Cotrino, C.R. 2007. Removal of Hydrogen Sulfide from Groundwater Using Packed-Bed Anion Exchange Technology. *Fla. Water Resources Jour.*, November; 22–25.

Dell'Orco, M., et al. 1998. Sulfide-Oxidizing Bacteria: Their Role During Air Stripping. *Jour. AWWA*, 90:107.

Lyn, T. L. and J.S. Taylor. 1991. Assessing Sulfur Turbidity Formation Following Chlorination of Hydrogen Sulfide in Groundwater. *Jour. AWWA*, 84:103.

Monscvitz, J.T. and L.D. Ainsworth. 1974. Treatment for Hydrogen Polysulfide. *Jour. AWWA*, 66:537.

Sammons, L.L. 1959. Removal of Hydrogen Sulfide from a Ground Water Supply. *Jour. AWWA*, 51:1275.

CHAPTER EIGHT

Nitrate Removal

Sources of nitrate in groundwater include nitrogen (added as an inorganic fertilizer), animal manure, fossil fuel combustion, lawn fertilizers, septic systems, and domestic animals in residential areas. Nitrate can persist in groundwater for long periods of time, and levels can increase over time with increased loading. Areas with a high risk of groundwater contamination generally have high nitrogen loading or high population density, well-drained soils, and less extensive woodland relative to cropland. Depth of groundwater also plays an important role in nitrate concentrations. Ingestion of drinking water containing nitrate by infants can cause low oxygen levels in the blood. The US Environmental Protection Agency (USEPA) has set a maximum contaminant level of 10 mg/L as nitrogen for nitrate.

TREATMENT ALTERNATIVES

Four treatment processes are generally considered acceptable for nitrate removal. These are anion exchange, biological removal, membrane filtration, and electrodialysis. Currently, anion exchange is used most frequently because of lower capital and operating costs. However, use of biological nitrate removal is expected to increase as more commercially viable systems are developed and stricter controls are placed on brine discharges. Biological removal is a promising but still developing technology for groundwater wellhead applications. Membrane filtration using reverse osmosis or nanofiltration and electrodialysis may be most useful when the water has high levels of sulfate, chloride, or total dissolved solids.

Anion Exchange

Anion exchange is a relatively simple, moderately priced alternative for nitrate removal. The process uses a resin to exchange nitrate for chloride at the anion-exchange resin surface, which is regenerated with brine solution periodically. The main drawback to this technology is the discharge of water with high levels of total dissolved solids (TDS), which can range from 10,000 to more than 100,000 mg/L, depending on the water quality and whether or not waste minimization modifications are used. Key water quality parameters are nitrate, arsenic, sulfate, bicarbonate alkalinity, chloride, and pH. Figure 8-1 shows a typical process flow diagram for anion exchange.

The efficiency of the anion-exchange process is affected by the quality of the water entering the system. The anion-exchange resin will remove sulfate and bicarbonate alkalinity, as well as other anions. As a result, the run length will be shortened if the raw water has high concentrations of sulfate and bicarbonate alkalinity or other anions.

The anion-exchange process for nitrate removal is similar to cation-exchange softening except that negatively charged monovalent ions are being removed and nitrate is not the most preferred common ion removed in the exchange unit. Standard and nitrate-selective chloride-form strong base anion (SBA) exchange resins are used for nitrate removal. Excess sodium chloride or calcium chloride at a concentration of 1.5 to 12% is used for regeneration.

The term *nitrate selective* refers to resins that retain nitrates more strongly than any other ions including sulfates. Nitrate-selective resins are similar to standard resins but have larger chemical groups on the nitrogen atom of the amine than the methyl groups that comprise a standard resin. The larger size of the amine groups makes it more difficult for divalent ions such as sulfates to attach to the resin. This reorders the affinity relationships so that nitrate has a higher affinity for the resin than sulfate, even at drinking water concentrations.

Selectivity for standard resins generally follows:

sulfate > nitrate > chloride > bicarbonate

Selectivity for nitrate selective resins is:

nitrate > sulfate > chloride > bicarbonate

Because all commercially available SBA standard resins prefer sulfate to nitrate at the TDS levels and ionic strengths of typical groundwater, chromatographic peaking of nitrate occurs following its breakthrough. Chromatographic peaking is the dumping of high concentrations of an ion from the resin bed as exchange sites are used up. In nitrate-selective resins, sulfate is the ion dumped from the resin after breakthrough.

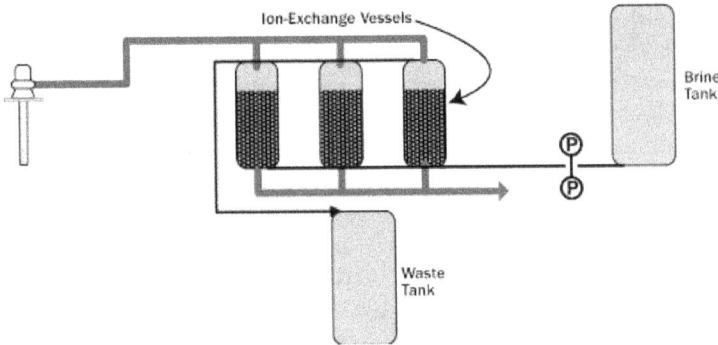

Figure 8-1 Process flow diagram for anion exchange
Courtesy of Paul Mueller, CH2M HILL

When peaking occurs, the effluent nitrate concentrations (in the case of standard resins) that exceed the source water nitrate concentration. Nitrate peaking depends primarily on the water quality, including TDS, sulfate, nitrate, and alkalinity concentrations, as well as the type of SBA resin used. In high TDS waters, nitrate may be preferred over sulfate even with standard resins.

Regeneration can be accomplished with stronger- or weaker-strength brine concentrations. Using weaker solutions may be more efficient in many cases but requires more frequent regeneration and can result in nitrate leakage.

Regeneration can be completed in cocurrent (downflow) or countercurrent (upflow) mode. Although countercurrent regeneration may be more efficient it requires stabilization of the bed so that it does not fluidize and mix during regeneration. Some vendors have developed sophisticated methods for stabilizing the bed in order to take advantage of countercurrent regeneration efficiencies, while others use completely full ion-exchange vessels, called packed bed systems, that cannot be fluidized. Systems are also available that have a mixed bed followed by a separation step to capture and recycle resin. A comparison of ion-exchange nitrate-removal systems is provided in Table 8-1.

Minimization of the waste stream has been the subject of innovation in ion-exchange systems. Many vendors reuse parts of the waste stream and discharge others. Typically a resin bed is backwashed, dosed with a brine solution, and then rinsed before being put back on line. By capturing and reusing portions of the rinse and regeneration waste streams, waste can be minimized. The amount of waste stream that can be reused for rinse recycle or reused in the brine stream depends heavily on the quality of the water being treated. Most vendors have models that will predict the amount of waste discharged. For example, in Glendale, Arizona, for water with averaged nitrate of 16 mg/L and sulfate of 100 mg/L, an alkalinity of 100 mg/L, and a raw water pH of 7.8, the discharge waste stream is approximately 0.5% of production. Compare this to a waste stream of 5% without waste minimization techniques.

Design considerations. Anion-exchange units are typically of the pressure type, downflow design. Automatic regeneration based on volume of water treated is normally considered in the design process. Multiple vessels must be provided so that at least one vessel is off-line for regeneration. Often, a portion of the water is bypassed

Table 8-1   Comparison of Nitrate-Removal Ion-Exchange Technologies

Type of System	Standard Bed With Cocurrent or Countercurrent Regeneration	Packed Bed With Cocurrent or Countercurrent Regeneration	Mixed Bed
Standard equipment provided	Brine tanks, regenerant pumps, ion-exchange vessels, valves, piping, resin, control system	Pretreatment filter, brine tanks, regenerant pumps, ion-exchange vessels, valves, piping, resin, control system	Brine tanks, regenerant pumps, ion-exchange reactor, mixer, media collection tank, recycle pump, valves, piping, resin, control system
Resin types	Standard or nitrate selective	Standard or nitrate selective	Proprietary, standard or nitrate selective
Typical waste amounts	5% or more	1% or less	Less than 1%
Typical regenerant use	Salt at 8–10 lb/cu ft	Salt at 8–10 lb/cu ft	Salt at 8–10 lb/cu ft
Waste minimization strategy and equipment	Can achieve less than 1% waste when reclaim and reuse portions of rinse and regenerant waste streams; requires additional tanks, pumps, and controls	Does not backwash; can include waste reclamation and reuse of regenerant by reusing portions of rinse and regenerant streams; requires additional tanks, pumps, and controls	Regenerates a percentage of media in off-line regeneration tank, minimizing waste stream

around the unit and blended with the treated water. The maximum blend ratio must be determined based on the highest anticipated raw water nitrate level. Anion-exchange media will remove both nitrates and sulfate from the water being treated.

The treatment flow rate typically does not exceed 8 gpm/sq ft of bed area. The backwash flow rate is usually 2 to 3 gpm/sq ft of bed area because the resin has a low specific gravity. A fast rinse, which is approximately equal to the service flow rate, is provided. Adequate freeboard must be provided to accommodate the unit's backwash flow rate, unless the system is designed as a packed bed system. An adequate underdrain and supporting gravel system, brine distribution equipment, and cross-connection control are all needed in the vessel design.

Many vendors supply this equipment, and a number of equipment suppliers have implemented process modifications to minimize waste. Although easily automated, these systems require routine operation and maintenance. The system can be designed using pressure vessels, thus eliminating the need for repumping. This process is easily adaptable for seasonal use.

Operational considerations. Whenever possible, the treated water nitrate level should be monitored using a continuous nitrate analyzer that is equipped with a high-nitrate–level alarm. If continuous monitoring and recording equipment is not provided, the finished water nitrate levels should be sampled and tested daily, preferably just prior to regeneration of the unit.

Prior to any discharge, the reviewing authority must be contacted for wastewater discharge limitations or National Pollutant Discharge Elimination System requirements. Prechlorination of resins should be avoided, because chlorine may damage the resins and produce nitrosamine compounds including NDMA (N-nitrosodimethylamine) that may have health implications.

Disposal of nitrate-contaminated brine. Because of its eutrophication potential, nitrate-contaminated brine usually cannot be disposed of into rivers or lakes, even if it is slowly metered into the receiving water. The high TDS and sodium concentration also prevent disposal of spent regenerant onto land where its nitrogen content could serve as a fertilizer. It is feasible to use potassium chloride as a regenerant, but it is more expensive than sodium chloride. Some researchers are looking at ways to precipitate the calcium carbonate in brine wastes and reuse the product in wallboard and other products, but no commercially viable application is currently in place.

Discharge to a sanitary sewer is possible in some places. Careful coordination is needed to evaluate the impact on the sewage treatment process; some locations limit the TDS concentration that can be discharged to the wastewater treatment plant.

Removing the nitrate from the spent brine prior to its reuse is possible via biological denitrification. Bench- and pilot-scale studies of this process have been reported by Van der Hoek et al. (1987) who found that biological denitrification is feasible if the level of sodium chloride is below about 15,000 mg/L.

Many systems dispose of the brine in evaporation lagoons, and careful design is required. Researchers are exploring reuse alternatives including salt marsh development, spray evaporation, solar pond development that generates excess heat, and membrane concentration of brine waste.

Biological Removal

Biological denitrification can be accomplished with either autotrophic (without oxygen) or heterotrophic bacteria (with oxygen). Most installations use heterotrophic bacteria, which also require that a carbon source (or electron donor) be added to the raw water. The electron donor source can be vinegar, ethanol, or sucrose. The process includes growing denitrifying bacteria on a fixed or fluidized bed and postfiltration to remove bacteria that is carried over from the bioreactor. Many systems are designed to be in open vessels that require repumping. A typical flow diagram for this process is shown as Figure 8-2. Only a few vendors currently supply equipment for biological denitrification for wellhead applications. This process requires an initial start-up period, which is necessary for growth of bacteria and after the system has been off-line for long periods of time. Because the start-up period can last 30 to 60 days, most systems are run continuously. Solids are produced in the bioreactor and from filter backwashing.

Biological nitrate removal is a common wastewater treatment process but has not been widely used in drinking water treatment. Research being conducted in Glendale, Arizona, and other locations may allow the near-term use of this technology. Biological removal of nitrate is completed using a combination of autotrophic and heterotrophic bacteria. A carbon source is required to complete the removal; ethanol or vinegar is often used in drinking water applications. Biological nitrate removal technology is likely to continue to develop as more and more utilities and regulatory agencies struggle with residuals handling for brine wastes.

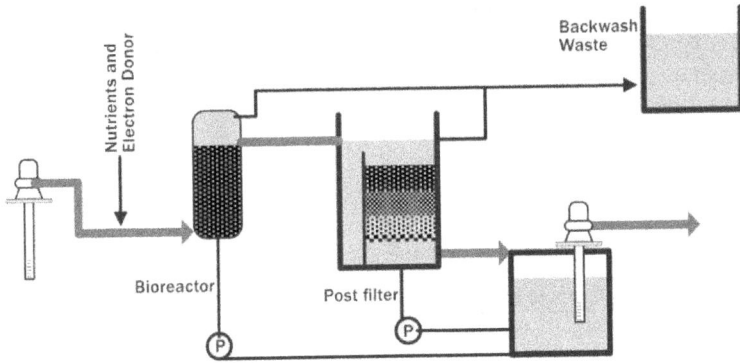

Figure 8-2 Biological denitrification process flow diagram
Courtesy of Paul Mueller, CH2M HILL

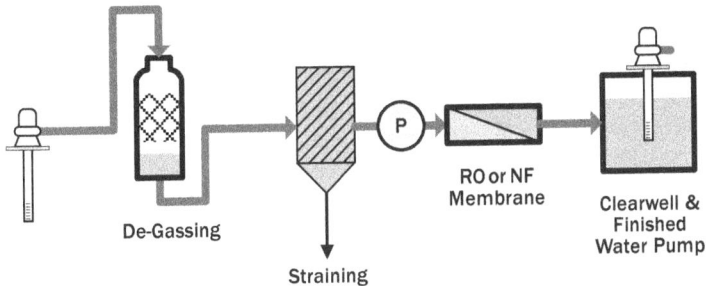

Figure 8-3 RO or NF membrane application with degassing and straining
Courtesy of Paul Mueller, CH2M HILL

Membranes and Electrodialysis Reversal

Reverse osmosis, nanofiltration, and electrodialysis reversal systems can effectively remove nitrate from water. However, as discussed previously, the high capital and operating costs of these systems generally limit their use to waters with other characteristics that require this treatment technology, such as very high TDS, saltwater intrusion, or radionuclide removal. Some applications have been used in groundwater wellhead applications where multiple treatment objectives, such as softening and nitrate or arsenic removal, would require a two-stage process. Figure 8-3 shows a typical process flow diagram for a simple reverse osmosis of nanofiltration application. Included in this diagram are pretreatment steps for degassing and straining. Degassing is important for high-pressure membrane applications where excess carbon dioxide or methane would result in membrane fouling.

Table 8-2 provides a comparison of treatment technologies for nitrate removal.

Table 8-2   Benefits and Drawbacks of Nitrate Treatment Alternatives

Nitrate Treatment Alternative	Benefits	Drawbacks
Anion exchange	Many commercially available systems Lowest capital cost Relatively easy to operate Easy to automate Also removes arsenic	High total dissolved solids liquid waste stream Efficiency is dependent on water quality
Biological removal	No brine waste	Requires postfiltration Few commercially available systems Requires carbon source and nutrients
Nanofiltration	Relatively easy to operate Also softens water and removes some inorganics and organics	May require extensive pretreatment Requires significant maintenance Operates at high pressure Relatively high capital and operating costs
Reverse osmosis	Relatively easy to operate Also softens water and removes many inorganics and organics	May require extensive pretreatment Requires significant maintenance Operates at high pressure Relatively high capital and operating costs
Electrodialysis reversal	Lower pressure requirements than other membrane systems Provides softening and removal of other inorganics and organics	May require extensive pretreatment

REFERENCES

Clifford, D., C. Lin, C., L. Horn, and J. Boegel. Nitrate Removal from Drinking Water in Glendale, Arizona. USEPA CR-807939, 1986.

Fox, K.R. Removal of Inorganic Contaminants from Drinking Water by Reverse Osmosis. USEPA.

Guter, G.A. Removal of Nitrate from Contaminated Water Supplies for Public Use. USEPA Grant No. R-805900-01-02-03.

Hubbard, R.K. and J.M. Sheridan. 1989. Nitrate Movement to Groundwater in the Southeastern Coastal Plain. *Jour. Soil and Water Conservation*, 44: January–February: 20.

Lowrance, R. 1992. Groundwater Nitrate and Denitrification in a Coastal Plain Riparian Forest. *Jour. Environmental Quality*, 21:July–September:401.

Meyer, Kerry. 2009, Pilot Testing of Nitrate-Contaminated Water: Ion-Exchange vs. Biological Treatment, Water Quality Technology Conference, Seattle.

Mueller, D. K., P.A. Hamilton, D.R. Helsel, K.J. Hitt, and B.C. Ruddy. 1995. Nutrients in Ground Water and Surface Water of the United States— An Analysis of Data Through 1992. US Geological Survey Water Resources Investigations Report 95-4031.

National Drinking Water Clearinghouse. 1997. Ion Exchange and Demineralization. Factsheet. 1997.

Nolan, B.T., B.C. Ruddy, K.J. Hitt, and D.R. Helsel. 1997. Risk of Nitrate in Groundwaters of the United States—A National Perspective. *Environmental Science and Technology*, 31:August:2229.

Solley, W.B., R.R. Pierce, and H.A. Perlman. 1993. Estimated Use of Water in the United States in 1990. US Geological Survey Circular 1081.

Spalding, R.F. and M.E. Exner. 1993. Occurrence of Nitrate in Groundwater—A Review. *Jour. Environmental Quality*, 22:July–September:392.

USEPA (US Environmental Protection Agency). 1995. Drinking Water Regulations and Health Advisories. Office of Water, Washington, D.C.

Van der Hoek, J.P, P.J.M. Latour, and A. Klapwijk. 1987. Denitrification with Methanol in the Presence of High Salt Concentrations and High pH. *Applied Microbiolical Biotechnology*, 27:199.

CHAPTER NINE

Uranium Removal

Uranium, a weak radioactive metal, occurs in the environment naturally and emits ionizing radiation due to radioactive decay. The health impact of high levels of uranium is unknown. Ingesting uranium causes kidney damage, which reduces the kidneys' ability to filter toxins from the bloodstream.

In the United States, uranium is predominately found in groundwater in the mountainous areas of the West. It is found in concentrated amounts in granite, metamorphic rocks, lignites, monazite sand, and phosphate deposits, as well as in the uranium-rich minerals of uraninite, carnotite, and pitchblende. Uranium must be oxidized before it is transported into groundwater. Once in solution, it remains for long periods of time.

Uranium typically exists in water as the uranyl ion, $(UO_2)^{+2}$, which formed in the presence of oxygen. At pH above 6, uranium exists in potable water primarily as the uranyl carbonate complex. This carbonate complex affects the efficiency of several treatment processes.

Uranium levels in laboratory tests are reported typically as micrograms per liter. In order to convert micrograms per liter to picocuries per liter, a ratio of U-234/U-238 of 0.68 to 1.3 is typically used. In California, it has been suggested that a conversion factor of 0.79 pCi/µg is more appropriate based on the alpha radiation activity in uranium isotopes found there. About 1% of US public water systems exceed the maximum contaminant level of 30 pCi/L; most of these are relatively small systems.

TREATMENT ALTERNATIVES

Technologies available for uranium treatment include enhanced coagulation–filtration ion exchange, lime softening, reverse osmosis, nanofiltration, activated alumina, and. Zero-valence iron media has been demonstrated at pilot scale to remove uranium.

Removal efficiencies achievable for each treatment alternative have been reported in the literature as

- coagulation/filtration: 80 to 95% (only at low pH or high pH)
- lime softening: 85 to 99%
- anion exchange: 90 to 100%
- reverse osmosis: 90 to 99%
- nanofiltration: 95%
- activated alumina: 90%

Coagulation/Filtration

Coagulation and filtration must be carried out within a narrow pH range. Coagulation at pH 6 or 10 typically removes 70 to 90% of uranium with ferric chloride and 50 to 80% with alum. At pH 4 and 8, little removal of uranium is achieved.

Ion Exchange

Ion exchange involves either a cation- or anion-exchange resin to remove uranium. Ion-exchange media can consist of naturally occurring materials, such as zeolite, or man-made resins. Ion exchange removes contaminants by moving a cation or anion (e.g., sodium or chloride) on the surface of the resin into the liquid phase. The relative order of affinity of strong base anion resins for some common ions in drinking water show uranium as the most preferred anion for exchange:

Uranium/Perchlorate >> Sulfate/Chromium > Selenium/ Arsenate > Nitrate > Chloride > Bicarbonate > Fluoride

Cation resin in the hydrogen form has been found to remove uranium, probably by converting the uranium complex to the uranium cation. Removal rates are in the 90 to 95% range, but the effluent pH will be low (about 2.5 to 3.5) and the resin used in this method is not selective, removing all cations.

Cation resin in the sodium form, operating as a softener, has limited use in uranium removal and is very dependent on pH. At pH 8.2, no uranium is removed, and at pH 5.6, there is about 70% removal. As the resin exhausts to the calcium form, removal is even less effective, with no removal at pH 8.2 or 7, some removal beginning to occur at pH 5.6, and 60% removal at pH 4.

Anion resin in the chloride form can easily reduce uranium levels by more than 90%. It can be used in a regenerable process or once-through. Regeneration of anion resins for uranium removal requires more concentrated brine than that used for nitrate or arsenic removal. Brine concentrations of 10 to 20% improve regeneration efficiency. Anion exchange works best at pH between 5.6 and 8.2. Above pH 8.2, uranium carbonate can precipitate, and at pH below 5.6, removal is less than 50%. Because changes in pH with ion exchange can dump uranium from the resin, pH should remain steady in a system treating for uranium.

Regeneration is needed and provided with chloride or hydroxide solutions, most often sodium chloride or sodium hydroxide. The spent regenerant solution containing the uranium must be disposed of properly. Of particular concern is whether the uranium has been concentrated sufficiently for the waste to be classified as a low-level radioactive waste. Often, utilities simply replace the resin when the concentration approaches the radioactive limit at which it can be disposed of in a municipal landfill.

Reverse Osmosis and Nanofiltration

Reverse osmosis (RO) and nanofiltration (NF) use semipermeable membranes to strain uranium carbonate compounds out of water. Several RO and NF membrane types have been tested, and all show better than 90% removal efficiency. RO and NF may require significant pretreatment, operation, and maintenance. The concentrate produced from the plant will have elevated uranium levels, as well as minerals and elevated total dissolved solids levels. Levels should be about two to five times the raw water concentrations in most applications.

Lime Softening

Lime softening does not require highly elevated pH for effective uranium removal. Lime upflow clarifiers operated above pH 10 can achieve greater than 80% removal efficiencies.

Activated Alumina

Activated alumina requires considerable operator attention. Also, competing anion concentrations may affect regeneration frequency. Other considerations include disposal issues and handling of regeneration chemicals (caustic soda and acid).

A comparison of treatment technologies for uranium removal is provided in Table 9-1.

Table 9-1   Benefits and Drawbacks of Uranium Treatment Alternatives

Treatment Alternative	Benefits	Drawbacks
Anion exchange	Many commercially available systems Lowest capital cost Relatively easy to operate Easy to automate Also removes arsenic	High total dissolved solids in liquid waste stream Efficiency is dependent on water quality
Cation exchange	Many commercially available systems Lowest capital cost Relatively easy to operate Easy to automate Also removes arsenic	High total dissolved solids in liquid waste stream Efficiency is limited above pH 8
Nanofiltration	Relatively easy to operate Also softens water and removes some inorganics and organics	May require extensive pretreatment Requires significant maintenance Operates at high pressure Relatively high capital and operating costs
Reverse osmosis	Relatively easy to operate Also softens water and removes many inorganics and organics	May require extensive pretreatment Requires significant maintenance Operates at high pressure Relatively high capital and operating costs
Lime softening	Provides softening and removal of other inorganics and organics	Requires significant operational oversight Requires frequent maintenance
Activated alumina	Moderate cost	Sensitive to water quality Requires regeneration with hazardous chemicals

RESIDUALS HANDLING

Treating water for naturally occurring uranium results in residual streams that are classified as "technologically enhanced naturally occurring radioactive materials" (TENORM). Numerous regulations govern the disposal of waste streams containing radionuclides, although there are no federal regulations specifically for TENORM. The following regulations could apply to water treatment plant residuals containing uranium.

The Resource Conservation and Recovery Act (RCRA; 40 CFR 239 and 282) establishes programs and standards for regulating nonhazardous solid waste under Subtitle D, hazardous wastes under Subtitle C, and underground storage tanks under Subtitle I. Municipal solid waste landfills (MSWLF) can accept commercial and industrial wastes passing paint filter test (i.e., no standing water) and toxicity characteristic leaching procedure testing. Sites that accept hazardous wastes include landfills, surface impoundments, waste piles, land treatment units, and underground injection wells and are subject to strict design and operating standards in 40 CFR 264 and 265.

The Clean Water Act establishes requirements for direct discharges of liquid waste and the discharge of liquid wastes to publically owned treatment works.

The Safe Drinking Water Act includes requirements that US Environmental Protection Agency (USEPA) develop standards for underground injection control to prevent future contamination of drinking water.

The Atomic Energy Act (AEA) requires the Nuclear Regulatory Commission (NRC) to regulate civilian commercial, industrial, academic, and medical use of nuclear materials. States (Agreement States) can enter into agreements to establish radiation protection programs under the NRC. The current list of Agreement States and contacts can be found at http://nrc-stp.ornl.gov/asdirectory.html.

The Department of Transportation (DOT) regulations (40 CFR 171 and 180) govern the shipping, labeling, and transport of hazardous and radionuclide materials.

The Comprehensive Environmental Response, Compensation, and Liability Act (CERCLA) applies to the release or threat of release of hazardous substances including radionuclides, which may endanger human health and the environment.

The presence of radioactivity does not make a waste hazardous, although removal of other substances along with radionuclides (such as arsenic) could.

The Low-Level Radioactive Waste Policy Act (42 USC 2021b(9)) defines low-level radioactive wastes. The definition includes source materials and by-product materials. Water treatment plant residuals do not fall within the definition of by-product materials, but uranium is included in the source materials listed. If the uranium concentration is below the limit defined as an "unimportant quantity," then the waste is exempt from NRC and Agreement State regulation. The limit for uranium for an unimportant quantity is 0.05% by weight (or approximately 335 pCi/g) for solids materials. If a waste has a higher concentration and has a total of no more than 15 lb total of radioactive material (0.05% of uranium in 30,000 lb of media would be 15 lb of uranium), then the waste is classified as a small quantity. Systems may not posses more than 150 lb of small-quantity radionuclide waste in one calendar year.

Decision trees have been developed by USEPA to help provide guidance on disposal of radioactive waste with elevated levels of TENORM. Decision Tree 1 applies to Solids Residuals Disposal and Decision Tree 2 applies to liquid residuals disposal (USEPA, 2002).

REFERENCES

Aieta, E.M. et al. 1987. Radionuclides in Drinking Water: An Overview. *Jour. AWWA,* 74:4:144.

Clifford, D. and Z. Zhang. 1994. Modifying Ion Exchange for Combined Removal of Uranium and Radium. *Jour. AWWA,* 86:4:214.

Cothern, C.R. and W.L. Lappenbush. 1983. Occurrence of Uranium in Drinking Water in the U.S. *Health Physics,* 45:89.

Farrel, J. et al. 1999. Uranium Removal from Groundwater Using Zero Valent Iron Media. *Groundwater,* 37:4:618.

Sorg, T.J. 1988. Methods for Removing Radium from Drinking Water. *Jour. AWWA,* 80:7:105.

USEPA (US Environmental Protection Agency). 2002. Implementation Guidance for Radionuclides. USEPA (4606M) EPA #816-F-00-002.

———. 2005. A Regulator's Guide to Management of Radioactive Residuals from Drinking Water Treatment Technologies. EPA #816-R-05-004.

CHAPTER TEN

Radium and Gross Alpha Removal

Radium, a weak radioactive metal, occurs in the environment naturally and emits ionizing radiation due to radioactive decay. The health effects associated with ingestion of water containing elevated radium levels may involve the ionization of body cells, leading to developmental abnormalities, cancer, or death. The lungs, myeloid stem cells, and bones of humans are particularly sensitive to this type of exposure.

All people are chronically exposed to background levels of radiation present in the environment. The probability of radiation-caused cancer or genetic effects is related to the total amount of radiation accumulated by an individual. At very low exposure levels, such as concentrations in drinking water that are below the MCL, the risks are very small and uncertain. The health risk models used by the US Environmental Protection Agency (USEPA) in setting drinking water standards assume that any exposure may be harmful. Radium-226 is primarily an alpha particle emitter, and radium-228 is primarily a beta particle emitter. The maximum contaminant level (MCL) for radium-226 plus radium-228 is 5 pCi/L.

Public drinking water systems with elevated levels of radium have been identified by the monitoring provisions required by the Safe Drinking Water Act. There are more than 500 public water systems in the United States with total radium concentrations that exceed the MCL of 5 pCi/L.

TREATMENT ALTERNATIVES

A number of treatment technologies can be used to remove radium from groundwater. For large systems, the Radionuclides Rule (2000) lists the following as best available technology (BAT): ion exchange, reverse osmosis (RO), and lime softening. Hydrous manganese oxides have also been used widely to remove radium. The iron and manganese removal processes can remove radium if manganese is present in the raw water. Hydrous manganese oxide is effective for radium removal; radium coprecipitates with manganese in this process.

Figure 10-1 Permanganate solution from the tanks above are combined with manganous sulfate to make freshly precipitated hydrous manganese oxide.

Typical removal efficiencies for common radium removal technologies are as follows:
- RO or EDR: 90 to 99%
- lime-soda ash softening: 80 to 95%
- cation exchange 65 to 95%
- hydrous manganese oxides: 50 to 90%
- aeration and iron removal: 12 to 38%

Hydrous Manganese Oxide

Initially used in very high doses to remove high radium concentrations from uranium mining wastes, the process of sorption onto freshly precipitated hydrous manganese oxides (HMO) has been adapted to remove radium from drinking water (Figure 10-1). Precipitated manganese dioxide is added to the water and then filtered out on a media filter. Several installations in Iowa, Illinois, and Minnesota exist; removal at these facilities ranges from 50 to 90%. Removal depends on a number of factors, including
- HMO dose. In tests, doses were 0.2, 0.5, and 1.0 mg/L (Valentine et al, 1990).
- Raw water radium concentrations: These concentrations ranged from 5 pCi/L to more than 150 pCi/L.

Figure 10-2 Aerators and mixing in hydrous manganese oxide solution tank are used to keep the precipitated manganese in solution prior to injection into the raw water for radium removal.

In addition to the HMO dose, time is needed for the filter media to reach equilibrium with the radium concentration after HMO is added to the raw water. Where HMO is added in front of an existing filter, radium initially desorbs from the media. After equilibrium is reached in 30 to 60 days, effluent radium levels are at steady state. If HMO feed is halted, the media will continue to remove radium for up to 30 days, as it again reaches equilibrium with the higher raw water radium concentrations. Where new filter media is used, radium desorption on start-up is not an issue.

HMO is produced on site by mixing manganous sulfate and potassium permanganate with 10% additional permanganate. The mixture forms a precipitated manganese oxide that must be kept in suspension by mixing (Figure 10-2). Chemical feed equipment must be compatible with high solids concentrations and strong oxidants. Large peristaltic (hose) pumps and carrier water systems that prevent feed line clogging are recommended.

Full-scale installations are operating with several different filter media including greensand, pyrolusite, sand, and dual media. Because the filter system is designed to remove precipitated manganese, filter loading rates are generally low or deeper media beds are used.

Ion Exchange

Radium is removed with a cation-exchange system that uses standard softening resins. Radium, which is preferentially removed before calcium and magnesium, is not dumped when the calcium and magnesium break through. Regeneration can be accomplished with brine, although a higher concentration than normally used for softening, typically 10 to 20%, is needed to remove the radium. Ion-exchange systems used for radium removal are operated as softeners. However, radium builds up on the resin bed and must be carefully monitored and properly disposed of when radium levels get too high.

Reverse Osmosis

RO is a physical process in which high pressure is used to force water through a semipermeable membrane, which cannot pass metals and salts. RO membranes reject ions based on size and electrical charge. The raw water is typically called feed; the product water is called permeate; and the concentrated reject is called concentrate. Common RO membrane materials include asymmetric cellulose acetate and polyamide thin-film composite. Common membrane construction includes spiral-wound or hollow fine fiber. Each material and construction method has specific benefits and limitations depending on the raw water characteristics and pretreatment.

A typical large RO installation includes a high-pressure feed pump; parallel first and second stage membrane elements (in pressure vessels); valving; and feed, permeate, and concentrate piping. All materials and construction methods require regular maintenance. Factors influencing membrane selection are cost, recovery, rejection, raw water characteristics, and pretreatment. Factors influencing performance are raw water characteristics, pressure, temperature, and regular monitoring and maintenance.

RO requires a careful review of raw water characteristics, and pretreatment must prevent membranes from fouling and scaling. It is necessary to remove suspended solids to prevent colloidal and biofouling, while removal of dissolved solids is necessary to prevent scaling and chemical attack. Large-installation pretreatment can include media filters to remove suspended particles; ion-exchange softening or antiscalant to remove hardness; temperature and pH adjustment to maintain efficiency; acid to prevent scaling and membrane damage; activated carbon or bisulfite to remove chlorine (postdisinfection may be required); and cartridge (micro) filters to remove some dissolved particles and any remaining suspended particles.

The operator must monitor the rejection percentage to ensure radium removal to below the MCL. In addition, regular monitoring of membrane performance is necessary to determine fouling, scaling, or other membrane degradation; use of trends to track membrane performance is recommended. Acidic or caustic solutions are regularly flushed through the system at high volume and low pressure with a cleaning agent to remove fouling and scaling, and the system is returned to service. RO stages are cleaned sequentially. Frequency of membrane replacement depends on raw water characteristics, pretreatment, and maintenance.

Lime Softening

Lime softening uses chemical addition followed by an upflow solids contact clarifier to accomplish precipitation and clarification. Lime and soda ash are added in sufficient quantities to raise the pH while keeping the levels of alkalinity relatively low in order to precipitate carbonate hardness.

Precipitation of calcium carbonate and radium and final clarification occur in the upflow clarifier. Water flows up and over the weirs, while the settled particles are periodically removed from the bottom of the clarifier.

Jar tests to determine optimum pH and alkalinity for coagulation, and resulting pH and alkalinity adjustment, may be required. Optimum pH for radium removal is usually 10.5, although if manganese is present, pH may be as high as 11.5 for effective radium removal.

Chemical feed equipment must be checked several times during each work period to prevent clogging and equipment wear and to ensure adequate chemical supply. All chemical feed systems, valves, and piping must be regularly checked and cleaned to prevent buildup of carbonate scale, which can cause plugging and malfunction. Similar procedures also apply to the sludge disposal return system, which takes the settled sludge from the bottom of the clarifier and conveys it to the dewatering and disposal processes.

Electrodialysis Reversal

Electrodialysis reversal (EDR) is an electrochemical process in which ions migrate through ion-selective semipermeable membranes as a result of their attraction to two electrically charged electrodes. A typical EDR system includes a membrane stack with a number of cell pairs, each consisting of a cation transfer membrane, a demineralized flow spacer, an anion transfer membrane, and a concentrate flow spacer. Electrode compartments are at opposite ends of the stack. The influent feedwater (chemically treated to prevent precipitation) and concentrated reject flow in parallel across the membranes and through the demineralized and concentrate flow spacers, respectively. The electrodes are continually flushed to reduce fouling or scaling. Careful consideration of flush feedwater is required.

Typically, the membranes are cation- or anion-exchange resins cast in sheet form; the spacers are high-density polyethylene; and the electrodes are inert metal. EDR stacks are tank-contained and often staged. Membrane selection is based on careful review of raw water characteristics. A single-stage EDR system usually removes 50% of the total dissolved solids (TDS); therefore, for water with more than 1,000 mg/L TDS, blending with higher-quality water or a second stage is required to reach a level of 500 mg/L TDS. EDR uses the technique of regularly reversing the polarity of the electrodes, thereby freeing accumulated ions on the membrane surface. Although this process requires additional plumbing and electrical controls, it increases membrane life, does not require added chemicals, and simplifies cleaning.

Typically, EDR systems for radium removal include pretreatment with antiscalant, acid addition for pH adjustment, and a cartridge filter for prefiltration. EDR membranes are durable and can tolerate pH from 1 to 10 and temperatures to 115°F for cleaning. EDR membranes can be removed from the unit and scrubbed; solids are generally washed off by turning the power off and letting water circulate through the stack. Electrode washes flush out by-products of the electrode reaction. These by-products are hydrogen, formed in the cathode spacer, and oxygen and chlorine gas, formed in the anode spacer. If the chlorine is not removed, toxic chlorine gas may form.

Depending on raw water characteristics and radium concentrations, the membranes will require regular maintenance or replacement. EDR requires system flushes at high volume and low pressure and reversing the polarity on the membranes for cleaning. Continuous flushing is required to clean electrodes. If used, pretreatment filter replacement and backwashing will also be required. The EDR stack must be disassembled, mechanically cleaned, and reassembled at regular intervals.

Table 10-1 provides a comparison of treatment alternatives for radium removal.

Table 10-1 Benefits and Drawbacks of Radium Treatment Alternatives

Treatment Alternative	Benefits	Drawbacks
Cation exchange	Many commercially available systems Lowest capital cost Relatively easy to operate Easy to automate Also removes calcium and magnesium	High total dissolve solids liquid waste stream Efficiency is dependent on water quality, especially sulfate High brine concentration is needed for regeneration to remove radium Produces liquid brine stream with elevated radium levels
Lime softening	Also softens water and removes some inorganics and organics	Requires significant operations and maintenance Relatively high capital and operating costs Produces sludge with elevated radium levels
Reverse osmosis	Relatively easy to operate Also softens water and removes many inorganics and organics	May require extensive pretreatment Requires significant maintenance Operates at high pressure Relatively high capital and operating costs Produces liquid brine stream with elevated radium levels
Hydrous manganese oxide	Low-cost alternative for radium removal, especially in systems with existing filters Works with many filter removal media	Requires monitoring and operations oversight Made on site and must remain mixed Careful design of chemical feed systems is needed Produces radium-concentrated backwash water

RESIDUALS HANDLING

Treating water for naturally occurring radium results in residual streams that are classified as "technologically enhanced naturally occurring radioactive materials" (TENORM). Numerous regulations govern the disposal of waste streams containing radionuclides, although there are no federal regulations specifically for TENORM. The following regulations could apply to water treatment plant residuals containing radium.

The Resource Conservation and Recovery Act (RCRA; 40 CFR 239 and 282) establishes programs and standards for regulating nonhazardous solid waste under Subtitle D, hazardous wastes under Subtitle C, and underground storage tanks under Subtitle I. Municipal solid waste landfills (MSWLF) can accept commercial and industrial wastes passing paint filter test (i.e. no standing water) and toxicity characteristic leaching procedure testing. Sites that accept hazardous wastes include landfills, surface impoundments, waste piles, land treatment units, and underground injection wells and are subject to strict design and operating standards in 40 CFR 264 and 265.

The Clean Water Act establishes requirements for direct discharges of liquid waste and the discharge of liquid wastes to publically owned treatment works (POTW).

The Safe Drinking Water Act includes requirements that USEPA develop standards for underground injection control to prevent future contamination of drinking water.

The Atomic Energy Act requires the Nuclear Regulatory Commission (NRC) to regulate civilian commercial, industrial, academic, and medical use of nuclear materials. States (Agreement States) can enter into agreements to establish radiation protection programs under the NRC. The current list of Agreement States and contacts can be found at http://nrc-stp.ornl.gov/asdirectory.html.

The Department of Transportation (DOT) regulations (40 CFR 171 and 180) govern the shipping, labeling, and transport of hazardous and radionuclide materials.

The Comprehensive Environmental Response, Compensation, and Liability Act (CERCLA) applies to the release or threat of release of hazardous substances including radionuclides, which may endanger human health and the environment.

The presence of radioactivity does not make a waste hazardous, although removal of other substances along with radionuclides (such as arsenic) could.

The Low-Level Radioactive Waste Policy Act (42 USC 2021b(9)) defines low-level radioactive wastes. The definition includes source materials and by-product materials. Water treatment plant residuals do not fall within the definition of by-product materials; radium is not included in the source materials listed (although uranium is).

Decision trees have been developed by USEPA to help provide guidance on disposal of radioactive waste with elevated levels of TENORM. Decision Tree 1 applies to Solids Residuals Disposal and Decision Tree 2 applies to liquid residuals disposal (USEPA, 2002).

REFERENCES

Dumbaugh, T. 2004. Hydrous Manganese Oxide (HMO) Process for Radium Reduction—Design and Practical Operating Experience. AWWA: Denver, CO.

Mott, H.V., S. Singh, and V.R. Kondapally. 1993. Factors Affecting Radium Removal Using Mixed Iron-Manganese Oxides. *Jour. AWWA*, 85:10:114.

USEPA (US Environmental Protection Agency). 2002. Implementation Guidance for Radionuclides. USEPA (4606M) EPA #816-F-00-002.

———. 2005. A Regulator's Guide to Management of Radioactive Residuals from Drinking Water Treatment Technologies. EPA #816-R-05-004.

Valentine, R.L. 1992. Radium Removal Using Preformed Hydrous Manganese Oxides. AWWA Research Foundation: Denver, Colo.

Valentine, R.L., K.M. Spangler, and J. Meyer. 1990. Removing Radium by Adding Preformed Hydrous Manganese Oxides. *Jour. AWWA*, 82:2:66.

CHAPTER ELEVEN

Barium Removal

Barium, a naturally occurring alkaline earth metal, is found primarily in the Midwest in combination with other chemicals such as sulfur or carbon and oxygen. Traces of the element are found in most surface waters and groundwaters. It can also be produced in oil and gas drilling muds, copper smelting, waste from coal-fired power plants, jet fuels, and automotive paints and accessories.

The health effects of barium in water differ for soluble and insoluble compounds. Barium compounds that do not dissolve well in water are not generally harmful and are often used for medical purposes. Water-soluble barium salt compounds that are toxic when ingested. The acetate, nitrate, and halide salts of barium are soluble in water, but the carbonate, chromate, fluoride, oxalate, phosphate, and sulfate salts are quite insoluble. The aqueous solubility of barium compounds increases as the pH decreases.

Short-term exposure above the maximum contaminant level (MCL) potentially causes gastrointestinal disturbances and nerve block, causing muscular weakness. Long-term exposures to barium at levels above the MCL have the potential to cause high blood pressure, changes in heart rhythm, brain swelling, and damage to the liver, kidney, heart, and spleen.

REMOVAL ALTERNATIVES

Potential treatment alternatives for barium removal and their reported achievable removal efficiencies are as follows:
- ion exchange: 93 to 98%
- reverse osmosis (RO): >90%
- lime softening: >90%
- electrodialysis reversal (EDR): >90%

Soluble barium removal with ion exchange is achieved using cationic resins in the chloride form. RO for soluble barium uses a semipermeable membrane operated under high pressure. Lime softening for soluble barium uses calcium hydroxide to raise pH above 10.5 and supersaturate the solution with calcium carbonate, which is then precipitated along with the soluble barium in the water. EDR uses ion-selective (cationic and anionic) membranes in which ions migrate through the membrane from a less concentrated solution to a more concentrated solution. Soluble barium is removed through the cationic membrane.

Barium sludge is typically dried and sent to a landfill after toxicity characteristic leaching procedure testing.

Ion Exchange

Cation-exchange resins works with standard softening resins. Barium is preferentially removed before calcium and magnesium and is not dumped when the calcium and magnesium break through. Regeneration can be accomplished with brine, although a higher concentration than normally used for softening is needed to remove the barium. Typically 10 to 20% brine solutions are used. If barium is not removed during regeneration, the resin can be soaked in a hydrochloric acid solution (10% solution); however, removal takes several hours to complete.

Weak acid cationic resins can also be used to remove barium, either in the sodium or hydrogen form. Regeneration of hydrogen resins is often accomplished with hydrochloric acid using a dose with approximately 20% excess acid above the theoretical amount needed.

Reverse Osmosis

RO is a physical process used to force water through a semipermeable membrane, which cannot pass metals and salts. RO membranes reject ions based on size and electrical charge. The raw water is typically called feed; the product water is called permeate; and the concentrated reject is called concentrate. Common RO membrane materials include asymmetric cellulose acetate or polyamide thin film composite. Common membrane construction includes spiral-wound or hollow fine fiber. Each material and construction method has specific benefits and limitations depending on the raw water characteristics and pretreatment.

A typical large RO installation includes a high-pressure feed pump; parallel first- and second-stage membrane elements (in pressure vessels); valving; and feed, permeate, and concentrate piping. All materials and construction methods require regular maintenance. Factors influencing membrane selection are cost, recovery, rejection, raw water characteristics, and pretreatment. Factors influencing performance are raw water characteristics, pressure, temperature, and regular monitoring and maintenance.

RO requires a careful review of raw water characteristics. In addition, pretreatment must prevent membranes from fouling and scaling. Suspended solids are removed in order to prevent colloidal and biofouling, and dissolved solids are removed in order to prevent scaling and chemical attack. Large-installation pretreatment can include media filters to remove suspended particles; ion-exchange softening or antiscalant to remove hardness; temperature and pH adjustment to maintain efficiency; acid to prevent scaling and membrane damage; activated carbon or bisulfite to remove chlorine (post-disinfection may be required); and cartridge (micro) filters to remove some dissolved particles and any remaining suspended particles.

The operator must monitor the rejection percentage to ensure barium removal to levels below the MCL. It is necessary to regularly monitor membrane performance in order to determine fouling, scaling, or other membrane degradation. Use of trends to track membrane performance is recommended. Acidic or caustic solutions are regularly flushed through the system at high volume and low pressure with a cleaning agent to remove fouling and scaling. The system is then flushed and returned to service; RO stages are cleaned sequentially. Frequency of membrane replacement depends on raw water characteristics, pretreatment, and maintenance.

Lime Softening

Lime softening uses a chemical addition followed by an upflow solids-contact clarifier to accomplish precipitation and clarification. Chemical addition includes adding lime and soda ash in sufficient quantities to raise the pH while keeping the levels of alkalinity relatively low in order to precipitate carbonate hardness.

Barium precipitates as $Ba(OH)_2$. Precipitation of calcium carbonate and barium hydroxide and final clarification occur in the upflow clarifier. The water flows up and over the weirs, while the settled particles are periodically removed from the bottom of the clarifier.

Jar tests to determine optimum pH and alkalinity for coagulation, and resulting pH and alkalinity adjustment, may be required. Optimum pH for barium removal is usually 10 to 10.5.

Chemical feed equipment should be checked several times during each work period to prevent clogging and equipment wear and to ensure adequate chemical supply. All chemical feed systems, valves, and piping must be regularly checked and cleaned to prevent buildup of carbonate scale, which can cause plugging and malfunction. Similar procedures apply to the sludge disposal return system, which takes the settled sludge from the bottom of the clarifier and conveys it to the dewatering and disposal processes.

Electrodialysis Reversal

EDR is an electrochemical process in which ions migrate through ion-selective semipermeable membranes as a result of their attraction to two electrically charged electrodes. A typical EDR system includes a membrane stack with a number of cell pairs, each consisting of a cation-transfer membrane, a demineralized flow spacer, an anion-transfer membrane, and a concentrate flow spacer. Electrode compartments are at opposite ends of the stack. The influent feedwater (chemically treated to prevent precipitation) and concentrated reject flow in parallel across the membranes and through the demineralized and concentrate flow spacers, respectively. The electrodes are continually flushed to reduce fouling or scaling. Careful consideration of flush feedwater is required. Dilute acid flush or dilute brine flushes are often used to reduce fouling, but manufacturers' recommendations should be followed.

Typically, the membranes are cation- or anion-exchange resins cast in sheet form; the spacers are high-density polyethylene; and the electrodes are inert metal. EDR stacks are tank-contained and often staged. Membrane selection is based on careful review of raw water characteristics. Because a single-stage EDR system usually removes 50% of the TDS, for water with more than 1,000 mg/L TDS, blending with higher-quality water or a second stage is required to achieve 500 mg/L TDS. EDR uses the technique of regularly reversing the polarity of the electrodes, thereby freeing accumulated ions on the membrane surface. Although this process requires additional plumbing and electrical controls, it does increases membrane life, does not require added chemicals, and eases cleaning.

Typically, EDR systems for barium removal include pretreatment with antiscalant, acid addition for pH adjustment, and a cartridge filter for prefiltration.

EDR membranes are durable and can tolerate pH from 1 to 10 and temperatures to 115°F for cleaning. They can be removed from the unit and scrubbed. Solids can be washed off by turning the power off and letting water circulate through the stack. Electrode washes flush out by-products of electrode reaction, which include hydrogen, formed in the cathode spacer, and oxygen and chlorine gas, formed in the anode spacer. If the chlorine is not removed, toxic chlorine gas may form.

Depending on raw water characteristics and barium concentrations, the membranes will require regular maintenance or replacement. EDR requires system flushes at high volume and low pressure, as well as reversing of the polarity on the membranes for cleaning. Flushing is continuously required to clean electrodes. If used, pretreatment filter replacement and backwashing will be required. The EDR stack must be disassembled, mechanically cleaned, and reassembled at regular intervals.

Table 11-1 provides a comparison of treatment alternatives for barium removal.

Table 11-1 Benefits and Drawbacks of Barium Treatment Alternatives

Treatment Alternative	Benefits	Drawbacks
Cation exchange	Many commercially available systems Lowest capital cost Relatively easy to operate Easy to automate Also removes calcium and magnesium	High total dissolved solids liquid waste stream Efficiency is dependent on water quality, especially sulfate High brine concentration is needed for regeneration to remove barium
Lime softening	Also softens water and removes some inorganics and organics	Requires significant operations and maintenance Relatively high capital and operating costs
Reverse osmosis	Relatively easy to operate Also softens water and removes many inorganics and organics	May require extensive pretreatment Requires significant maintenance Operates at high pressure Relatively high capital and operating costs
Electrodialysis reversal	Lower pressure requirements than for other membrane systems Provides softening and removal of other inorganics and organics	May require extensive pretreatment Relatively high capital and operating costs

REFERENCES

Cotton, F.A. and G. Wilkinson. 1980. *Advanced Inorganic Chemistry: Comprehensive Text*, 4th ed. New York: John Wiley, p. 286.

IPCS (International Programme on Chemical Safety). 2001. Barium and Barium Compounds. Geneva, Switzerland: World Health Organization, Concise International Chemical Assessment Document 33.

Lanciotti, E., et al. 1992. A Survey on Barium Contamination in Municipal Drinking Water of Tuscany. *Igiene Moderna*, 98:6:793.

Ohanian, E.V. and W.L. Lappenbusch. 1983. Problems Associated With Toxicological Evaluations of Barium and Chromium in Drinking Water. Washington, DC: US Environmental Protection Agency, Office of Drinking Water.

Subramanian, K.S. and J.C. Meranger. 1984. A Survey for Sodium, Potassium, Barium, Arsenic, and Selenium in Canadian Drinking Water Supplies. *Atomic Spectroscopy*, 5:34.

USEPA (US Environmental Protection Agency). 1984. Health Effects Assessment for Barium. Washington, DC: US Environmental Protection Agency.

———. 1985. Drinking Water Criteria Document for Barium. Washington, DC: US Environmental Protection Agency, Office of Drinking Water.

Willey, B.R. 1987. Finding Treatment Options for Inorganics. *Water/Engineering and Management*, 134:10:28.

CHAPTER TWELVE

Organic Compound Removal

Organic compounds may be natural or manmade in origin. Some compounds have individual regulated MCLs, while others may form regulated compounds after reaction with chlorine or other disinfectants. Others, like pharmaceutical and personal care products may not yet be regulated, but are of concern by consumers. Treatment technologies are often selected based on the classification of organic compound; synthetic, volatile, or natural. As more organic compounds become regulated, specifically pharmaceutical and personal care products, multiple treatment technologies may need to be employed to ensure effective removal.

The selection of a treatment technology for organic compounds is often complex and depends on many factors including the specific compound, or compounds, to be removed, concentration, water quality, site, and operational constraints. A list of potential treatment technologies was presented in Table 2-1. For organic compounds, these technologies included those listed in Table 12-1.

Table 12-1 Organic Compound Treatment Technologies

	Coagulation–filtration	Biological Filtration	Reverse Osmosis	Nanofiltration	Granular Activated Carbon	Anion Exchange	Electrodialysis Reversal	Excess Lime Softening	Aeration	Ozonation	Permanganate	Chlorine	Chlorine Dioxide	UV–Peroxide
Volatile organics					✓				✓	✓	✓	✓	✓	✓
Synthetic organics	✓	✓	✓	✓	✓	✓	✓			✓				✓
Pesticides	✓	✓	✓	✓	✓		✓	✓		✓				✓
Dissolved organic carbon	✓	✓	✓	✓	✓	✓	✓	✓		✓				✓
Disinfection by-product precursors	✓	✓	✓	✓	✓		✓	✓		✓				✓
Pharmaceuticals and personal care products	✓	✓	✓	✓	✓		✓	✓		✓		✓		✓

Figure 12-1   These packed-tower air strippers in Tacoma, Washington, are used to strip PCE from groundwater.

Thousands of individual organic compounds can be found in water. This chapter discusses some aspects of treatment technology selection for a few of the more common compounds.

Aeration

Volatile organic compounds, including trichloroethylene (TCE) and tetrachloroethylene (PCE), are most often removed using aeration technologies (Figure 12-1) (see Chapter 2). An important factor in determining how effective aeration will be is the Henry's constant for the specific compound or compounds to be treated. Other important considerations include water temperature (aeration is more difficult at lower temperatures), operational need to repump the water once it is exposed to atmospheric conditions, noise considerations from the aeration treatment, and, in many instances, the need to capture and treat the off-gas.

Typically, the following information is needed when evaluating aeration systems for treatment of organic compounds:

- groundwater flow rate,
- compound to be removed,
- influent concentration,
- maximum effluent concentration, and
- water temperature.

A required air-to-water ratio can be calculated with this information and different aeration technologies can be evaluated. Aeration technologies (Table 12-2) include the following types of aeration: spray, diffused, cascade, tray, and packed tower.

Table 12-2 Aeration Treatment Technologies

Aeration Type	Description	Relative Efficiency
Diffused	Bubbles run through a water column	Efficiency depends on amount and size of bubbles, height of water column, contact time
Spray	Spray nozzles in open or closed system	Efficiency depends on size of droplets formed, pressure, and ventilation
Cascade	Exposed slats or trays that rely on natural draft for air	Efficiency depends on height and number of trays, specific design for air induction
Slat-tray	Trays enclosed in a box with countercurrent airflow	Gas transfer and efficiency improved with forced-air induction, more trays, and higher box
Packed tower	Enclosed column filled with plastic media to break up water into small drops with countercurrent air flow	Can achieve very high levels of gas transfer, depending on height, type of media, and airflow

Off-gas is usually treated with granular activated carbon (GAC) gas-phase adsorption. Contactors are designed to adsorb the collected off-gas from the aeration technology when required. The efficiency of GAC adsorption in the off-gas, compared to that in water, is often much higher and, as a result, the contactors are often relatively small. In some cases, the off-gas can also be treated biologically using GAC contactors.

Biological Removal

Several organic compounds, including PCE, natural organic compounds (humic and fulvic compounds, dissolved organic carbon) and many synthetic organic compounds (some personal care products, pharmaceutical compounds, and methyl-tert-butyl-ether [MTBE]), have been shown to be reduced or removed by biological filtration. In most cases, the biological filtration system is similar to the type used for heterotrophic nitrate removal and includes oxygen and carbon source additions to a biologically active contactor, followed by posttreatment including filtration.

Reverse Osmosis, Electrodialysis Reversal, and Nanofiltration

Reverse osmosis and electrodialysis reversal are used to effectively remove most organic compounds including dissolved organic compounds, natural organic compounds, synthetic organic compounds, pesticides, disinfection by-product (DBP) precursors, and many pharmaceutical and personal care products. Nanofiltration is effective for many larger naturally occurring organic compounds. Membrane processes are not very effective at removing volatile compounds, which often must be removed in a posttreatment step of degassing or adsorption.

GAC Adsorption

Many organic compounds can be treated using GAC adsorption (Figure 12-2). There are different types of GAC, and some exhibit better adsorption characteristics than others. The first step in evaluating GAC treatment is to evaluate GAC adsorption isotherms. The adsorption isotherm is for a specific compound and a specific GAC type, although often more than one type of GAC is displayed on an isotherm. A typical adsorption isotherm, shown in Figure 12-3, provides a means to estimate how many milligrams of the organic compound can be removed for a particular equilibrium (effluent) concentration. Isotherms are generated at a specific water temperature.

For many compounds, an iodine number is used for adsorption capacity, which compares the adsorption of a specific compound to that of iodine. The iodine number can be used to size vessel capacity and predict how long the material in the vessel will adsorb the contaminant before it is exhausted. For difficult-to-adsorb compounds,

Figure 12-2   These vertical vessels hold GAC, which is used to remove PCE from the groundwater.

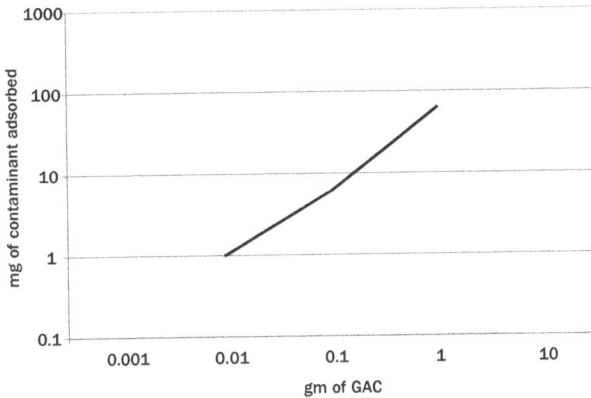

Figure 12-3 Example isotherm for adsorption capacity of GAC, mg/gm

such as MTBE, a trace capacity number (TCN) may be used. This number compares removal to that of acetotoxime and can also be used to predict the performance of these compounds. Isotherms, iodine numbers, and TCNs can be readily obtained from GAC vendors, and many are published in treatment texts.

Frequently used GAC sizes include are 8 × 30 US mesh and 12 × 40 US mesh (about 1 to 1.5 mm), although smaller, 30 × 50 US mesh (about 0.5 mm) is sometimes used to reduce the size of the contactor or extend the time to exhaustion. Typical contact times for GAC adsorption range from 5 to 20 min.

GAC is manufactured from bituminous coal or coconut shells and may be activated by re-agglomeration or direct activation. Re-agglomeration is the addition of man-made pores into the GAC structure before baking; direct activation is activation by baking without the addition of these pores. Re-agglomeration often results in a longer run time to exhaustion for two GACs with similar or identical iodine numbers.

When exhausted, GAC is often returned to the manufacturer for re-activation and replaced with either reactivated or virgin GAC. The decision to use re-activated GAC or virgin GAC is usually based on economics.

Anion Exchange

Anion exchange has been used effectively for many full-scale applications to remove naturally occurring organic compounds, including

Figure 12-4   Hydrogen peroxide feed can be combined with UV reactors like those shown above to oxidize organic contaminants.

DBP precursors. The anion-exchange systems, usually strong base resins in the chloride form, and are regenerated with a brine solution. The systems are similar to those used for nitrate removal (see Chapter 8).

Lime Softening

Lime softening removes some total organic compounds and has been demonstrated to remove many pharmaceutical and personal care compounds, as well as DBP precursors. Lime softening has also been shown to remove organic compounds added in polymers as coagulants or coagulant aids.

Oxidants and Advanced Oxidants

Ozone, chlorine, chlorine dioxide, permanganate, and advanced oxidation processes such as ultraviolet light combined with peroxide and ozone combined with peroxide can oxidize specific organic compounds (Figure 12-4). However, all of these processes produce some form of oxidation by-product. Many texts and research projects present evaluations of specific and general types of organic compound oxidation. However, there is general agreement that stronger oxidation processes such as ozone and advanced oxidation are required to provide destruction of man-made organic compounds without resulting in by-products that are likely to have some health effects. Ozone is usually effective for oxidation of some micropollutants, phenolic compounds, amines, and dyes. Advanced oxidation is usually effective for broad classes of organic compounds including micropollutants, aromatics, aliphatic hydrocarbons, phenols, amines, chlorinated organic pesticides, and dyes.

REFERENCES

AWWA (American Water Works Association) 1999. *Water Quality and Treatment*, 5th ed. McGraw-Hill: New York.

———. 1999. *Water Treatment Plant Design*, 5th ed. McGraw-Hill: New York.

Bower, E.J. and P.L. McCarty. 1982. Removal of Trace Chlorinated Organic Compounds by Activated Carbon on Fixed-Film Bacteria. *Environmental Science and Technology*, 16:836.

Dobbs, R.A. and J.M. Cohen. 1980. Carbon Adsorption Isotherms for Toxic Organics. USEPA #600/8-80-023.

Love, O.T. and R.J. Miltner. 1985. Removal of Volatile Organic Contaminants From Ground Water by Adsorption. USEPA, Office of Research and Development.

Megonnell, N. and A. McClure. 2003. Applying Activated Carbon to MTBE Removal. *Water Technology*, 26:2.

Orme-Zavaleta, J. 1991. Drinking Water Health Advisory. USEPA. Office of Drinking Water.

Rakness, K.L. 2005. *Ozone in Drinking Water*. AWWA: Denver, Colo.

Appendix A
Materials Compatibility for
Chemical Feed Systems

Material

Chemical	304 Stainless	316 Stainless	Brass	Bronze	Cast Iron	Copper	Chlorinated Polyvinyl Chloride	Ductile Iron	Ethylene Propylene-Dimonomer	Epoxy	Hypalon	Kynar	Neoprene	Nylon	Polyethylene	Polypropylene	Polyvinyl Chloride	Rubber	Teflon	Titanium	Tygon
Acetic acid	A	A	A	A	C	D	A	C	A	A	C	A	C	A	A	A	A	C	A	A	D
Aluminum chloride	C	C	D	D	D	D	A	D	A	A	B	A	A	D	B	A	A	A	A	B	B
Alum	B	B	D	D	D	D	A	D	A	A	A	A	A	A	A	A	A	A	A	A	B
Aluminum sulfate	B	B	D	D	D	D	A	D	A	A	A	A	A	A	A	A	A	A	A	A	B
Ammonia, anhydrous	A	A	D	D	A	D	A	A	A	A	D	A	C	A	B	A	B	D	A	C	A
Ammonia, aqueous	B	A	D	D	A	D	A	A	A	A	D	A	C	B	C	A	A	D	A	A	B
Calcium hydroxide	C	B	D	D	A	D	A	A	A	A	A	A	A	A	B	A	A	A	A	A	—
Carbon dioxide	A	A	A	A	D	D	A	D	B	A	A	A	B	A	C	A	A	B	A	A	—
Chlorine gas	B	B	—	—	—	—	A	—	A	—	B	A	C	D	B	C	A	D	A	A	—
Chlorine dioxide	D	D	D	D	D	D	C	D	B	D	B	A	B	D	B	B	C	C	A	D	A
Ferric chloride	D	C	D	D	D	D	A	D	A	A	B	A	B	A	A	B	A	A	A	C	B
Ferric sulfate	B	B	D	D	D	D	A	D	A	A	B	A	B	A	A	B	A	A	A	A	B
Ferrous chloride	D	C	C	C	D	C	A	D	—	A	A	A	A	C	A	A	A	A	A	A	B
Ferrous sulfate	B	B	D	D	D	D	A	D	A	A	B	A	B	A	A	A	A	A	A	A	B
Fluorosilicic acid	C	B	D	D	D	D	A	D	A	A	A	A	A	D	A	A	A	A	A	A	B
Hydrochloric acid	D	D	D	D	D	D	A	D	C	C	A	A	D	D	B	C	A	A	A	D	B
Hydrogen peroxide	B	B	D	D	B	D	A	B	A	A	B	A	A	A	C	B	C	B	A	B	D

A = excellent; B = good; C = fair; D = not recommended; — = not used or no information

Material

Chemical	304 Stainless	316 Stainless	Brass	Bronze	Cast Iron	Copper	Chlorinated Poly-vinyl Chloride	Ductile Iron	Ethylene Propy-lene Dimonomer	Epoxy	Hypalon	Kynar	Neoprene	Nylon	Polyethylene	Polyproplyene	Polyvinyl Chloride	Rubber	Teflon	Titanium	Tygon
Lime	A	A	A	A	A	A	—	A	D	A	—	A	B	A	A	A	D	—	A	A	A
Magnesium hydroxide	B	A	A	A	A	A	A	A	A	A	A	A	B	A	A	A	A	B	A	A	C
Nitric acid	A	A	D	D	D	D	D	D	D	D	D	A	D	D	C	D	D	D	A	A	A
Ozone	B	A	—	B	—	A	A	—	A	—	A	—	C	D	C	B	B	D	—	—	—
Phosphoric acid	D	B	D	D	D	D	A	D	B	—	C	A	D	B	B	B	B	D	A	C	D
Potash	A	A	D	D	C	C	C	C	—	A	—	—	A	A	B	A	C	B	—	—	B
Pottasium permanganate	B	B	C	C	A	C	A	C	A	A	—	A	A	A	A	A	A	B	A	A	—
Sodium bicarbonate	A	A	A	A	C	A	A	C	A	A	D	A	B	A	A	A	A	A	A	A	B
Sodium carbonate	B	B	B	B	A	B	A	A	A	A	—	A	A	A	A	A	A	B	A	A	—
Sodium chloride	B	C	C	C	A	C	A	C	A	A	—	A	A	A	A	A	A	A	A	A	B
Sodium hypochlorite	D	D	D	D	D	D	C	D	B	D	B	A	B	D	B	B	C	C	A	C	A
Sodium hydroxide	D	D	D	D	D	D	A	D	B	B	A	A	B	A	B	A	A	A	A	B	C
Sodium phosphate	B	A	D	C	D	C	A	C	A	A	A	A	B	A	A	A	A	A	A	A	A
Sodium silicate	B	A	A	A	A	B	A	B	A	A	C	A	A	A	A	A	A	A	A	A	—
Sulfur dioxide	D	A	B	B	D	B	A	—	B	C	B	A	B	C	B	C	A	B	A	A	C
Sulfuric acid	D	D	D	D	D	D	C	D	B	C	B	A	D	D	B	C	A	D	A	D	D

A = excellent; B = good; C = fair; D = not recommended; — = not used or no information

Index

NOTE: *f.* indicates a figure; *t.* indicates a table.

199

About the Author

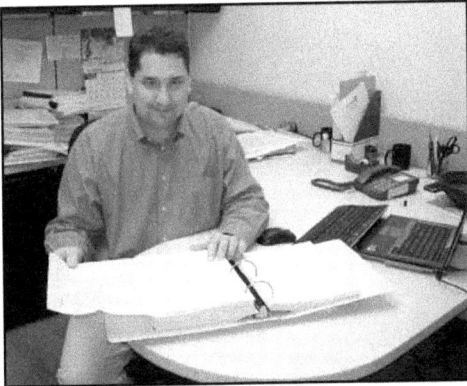

Lee Odell is a professional engineering consultant with CH2M HILL in Portland, Oregon. He received a bachelor's degree in civil engineering and a master's degree in civil and environmental engineering from the University of Iowa.

Lee has designed more than 100 groundwater treatment plants across the United States. In 1996 he designed the first municipal groundwater treatment plant to use manganese dioxide media for iron and manganese removal in Vancouver, Washington. In 1998 he designed the first full-scale water treatment plant in the United States to use granular ferric hydroxide for arsenic removal in Southern California.

Lee is an active AWWA member serving on national and local committees, including the Education Committee, Program Committee, Water Treatment Committee, and Water Quality Committee.

www.ingramcontent.com/pod-product-compliance
Lightning Source LLC
Chambersburg PA
CBHW052012230326
41598CB00078B/2981